The
GEOLOGICAL
EVOLUTION
of the
RIVER NILE

The
GEOLOGICAL
EVOLUTION
of the
RIVER NILE

RUSHDI SAID

Senior Research Scientist
Institute of Earth and Man
Southern Methodist University
Dallas, Texas

Senior Research Scientist
Conoco Oil Company

With Appendix C by
Felix P. Bentz and Judson B. Hughes
Santa Fe Minerals Inc., Dallas, Texas

With 73 Illustrations

Springer-Verlag
New York   Heidelberg   Berlin

Rushdi Said
Senior Research Scientist
Institute of Earth and Man
Southern Methodist University
Dallas, Texas 75275

Cover illustration: The Nile at Memphis.

Library of Congress Cataloging in Publication Data

Said, Rushdi.
  The geological evolution of the River Nile.

  Bibliography: p. 137
  Includes index.
  1. Geology—Nile Valley.   2. Paleogeography—Nile
Valley.   I.   Title.
QE328.S297   551.48′30962   81-5806

Additional material to this book can be downloaded from http://extras.springer.com

The use of general descriptive names, trade names, trademarks,
etc. in this publication, even if the former are not especially
identified, is not to be taken as a sign that such names, as understood by the Trade Marks and Merchandise Marks Act, may
accordingly be used freely by anyone.

9 8 7 6 5 4 3 2 1

ISBN 978-1-4612-5843-8   ISBN 978-1-4612-5841-4 (eBook)
DOI 10.1007/978-1-4612-5841-4

Preface

This book gives the geological history of the river Nile since it started to excavate its course in the Egyptian plateaus in late Miocene time in response to the lowering sea level of the desiccating Mediterranean. It formed a canyon longer, deeper, and just as awe inspiring as the Grand Canyon, Arizona. The canyon was transgressed by the advancing Mediterranean as it started filling during the early Pliocene, and since then by a number of rivers which ebbed and flowed as they succeeded one another. The modern Nile is a recent and humble successor to mighty rivers which once occupied the Nile Valley.

Dallas, Texas   Rushdi Said
August 1981

Acknowledgments

This book is based on field work carried out in Egypt during the seasons 1961–1978 while the author was a member of the Combined Prehistoric Expedition sponsored by Southern Methodist University, the Polish Academy of Science, and the Geological Survey of Egypt. Grateful acknowledgment is made to Professor Fred Wendorf, leader of the Expedition, and to several members for their fruitful discussions. Notable among these are Dr. Claude Albritton, Southern Methodist University, and Dr. J. De Heinzelin, University of Ghent, Belgium. The field work was aided by geologists M. S. Abdel Ghany and A. Zaghloul of the Geological Survey of Egypt. The drafting was by Reed Ellis and Hoda S. Armanious. I am also grateful to Dr. M. K. Ayouti, the General Manager of the Egyptian General Petroleum Corporation, for his help in providing data and samples of the deep boreholes of the delta and for his encouragement to bring this manuscript to publication. Dr. Fouad Y. Michael, Atlantic Richfield Oil Co., Dallas, Texas, drew my attention to many papers that were of pertinence to this work.

Dr. Felix P. Bentz and Judson B. Hughes, Santa Fe Minerals Inc., Dallas, Texas, graciously accepted to write Appendix C of this book which gives support, by the use of seismic data, to the theory here advanced on the origin of the valley of the Nile. Due acknowledgment is made to Santa Fe Minerals Inc. and its partners Bow Valley Exploration Ltd., Central Energy Development Co., and Citco International Petroleum Co. for permission to publish many of the seismic sections across the Nile in the area they held for oil exploration work in Egypt.

The preparation of this work for publication was made while the author occupied the position of senior research scientist, Institute of Earth and Man, Southern Methodist University. I am deeply indebted to Professor James E. Brooks, Provost of the University, for his continued help and encouragement. I owe a special word of thanks to Mr. Coy H. Squyres, President, Conoco Oil Co., Egypt, for his continued interest and support; not only was he generous with his time in countless discussion sessions, but also with the facilities of Conoco which greatly helped in the preparation of this book.

Due acknowledgment is given to the American Association of Petroleum Geologists, Roland Press, Academic Press, University of Chicago Press and Southern Methodist University Press for permission to publish some figures.

Contents

CHAPTER I

Introduction

This book presents the geological evolution of the River Nile within the boundaries of Egypt through a critical synthesis of a large amount of information worked out and assembled over many years. It aims at finding some order in the extremely complex stratigraphy of the fluvial and other sediments which fill the Nile Valley and which escaped erosion and/or diligent human interference. The younger sediments of the Nile are almost barren of datable materials except for the youngest of these which yielded, in addition to archeological materials, a reasonable number of radiocarbon dates and a few fossil remains. The older sediments of the Nile, now known through a large number of bore-holes, though not including materials amenable to radiometric measurements, are fossiliferous. Much of the stratigraphic order adopted in this book for the younger sediments is based on field relations as well as on correlations with sections outside Egypt, especially those described in detail from the Mediterranean and Red Sea areas where extensive oceanographic and submarine geological work has been recently carried out. Mechanical, mineral, and other physical characteristics of the different units were of help in correlating sediments; so were the archeological materials systematically collected from the most recent of the Nile sediments (Neonile) by the Combined Prehistoric Expedition. Correlations of the recently worked out stratigraphy of the Pleistocene, as it emerged from the study of the deep sea cores, with the classical systems of the land-based Pleistocene produced a valuable scale (Kukla, 1977).

Many seasons of field work aided in the working out of the stratigraphy of the deposits of the Nile Valley and their mapping. The accompanying maps and cross sections (Figures 1, 2, and 3) depict the distribution of the different units into which the deposits of the valley are subdivided. The large amount of well and borehole information (Appendix A) and reflection seismic data (Appendix C) proved invaluable instruments in formulating the theory in accordance with which the stratigraphy of the Nile and its geological history were worked out. The rapid expansion of population centers and the reclamation of lands along the edges of the valley and delta are fast changing the landscape of the country and destroying much of the geology of the valley. The accompanying maps, therefore, salvaged the geology of large tracts of land which held the last remains of the sediments of the Nile.

The valley and delta troughs of the Nile offered the only environment in Egypt which favored the accumulation and preservation of sediments during the Quaternary. Physical records of the presence of this epoch in other environments are sporadic and incomplete; the Quaternary was an epoch of intense erosion. Correlation of these records is difficult, and all classifications of these sediments are based on the relative or absolute elevations of existing and exhumed land surfaces, rock-cut platforms, pediments, and terraces. Attempts to establish a climatostratigraphic system of the Egyptian Quaternary based on sequences of geomorphic features with specific climatic signatures are complicated by the fact that Egypt was subjected to two different climatic models, the

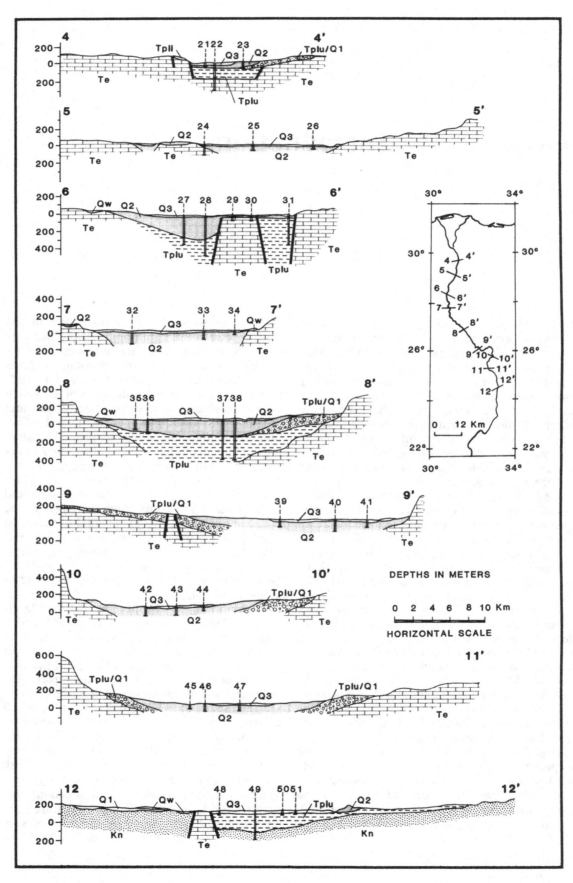

Figure 1. Cross sections in Nile Valley (see Table A-2 for location and stratigraphic data of boreholes).

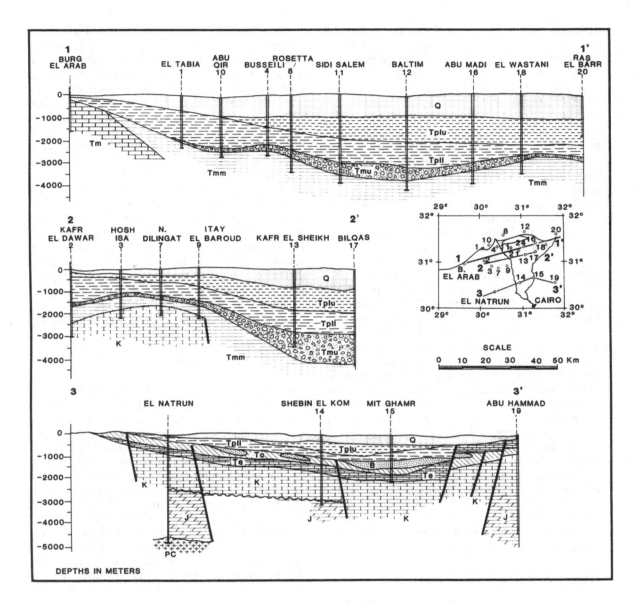

Figure 2. Cross sections in Nile delta (see Table A-1 for location and stratigraphic data of boreholes).

south by the northward migration of the Sudano-Sahelian savanna belt and the north by the Mediterranean pre-Sahara steppe. In addition to the evidence from the outcropping fluvial and other associated sediments, there is now an almost uninterrupted section of the Quaternary since the Nile deposits have been recently penetrated by deep drill holes. It is, therefore, appropriate to take the Nile as a standard section of the Quaternary of Egypt. Further work on the Quaternary of other environments is

Figure 3. Longitudinal section in Nile delta (see Table A-1 for location and stratigraphic data of boreholes).

planned, and it is hoped that the framework presented in this book will help bring order to the scattered records of this epoch in these environments.

In spite of the fact that the Nile has a very small discharge, it plays an important role in the life of Egypt. Flowing midway through the rainless wastes of the deserts of Egypt, the Nile provides Egypt with 98% of its water supply. The total amount of water carried by the river (averaging 86 billion cubic meters a year) makes the Nile one of the smallest rivers of the world in spite of its large basin and great length (Table I-1). The amount of water available to Egypt is 55.5 billion m^3/year which makes the share of each Egyptian 4 m^3/day.

The Nile is not only small relative to other rivers of the world, but it is also small compared to its predecessors. Since the incision of the valley of the Nile as a grand canyon, some $2^1/_2$ km deep, about 6,000,000 years ago, gigantic rivers have come and gone filling this canyon. These were succeeded by the last and extant Nile which entered Egypt hesitantly at the beginning of the Wurm glacial. Since then it has ebbed and flowed but without ever reaching the dimensions of its predecessors.

This book consists of three chapters and three appendices. Chapter 1 gives a broad outline of the Nile basin and the very tenuous relationship of the Egyptian Nile with its sources in equatorial Africa and Ethiopia affected by way of a series of cataracts across the Nubian swell. Chapter II describes the sediments of the successive rivers which occupied the valley since its incision within the Oligo-Miocene plateau of Egypt in late Miocene time.

It also describes the sediments which were formed during the intervals which separated the flow of these rivers. Chapter III deals with the geological history of the river in an attempt to give a time perspective to these sediments. Appendix A gives the basic data of the deep and shallow boreholes as well as the measured sections which were used in working out the stratigraphy of the deposits of the Nile and in constructing the cross sections across the valley and the delta (Figures 1–3). Appendix B lists the formational names used in this book, their type sections, lithology, stratigraphic limits, extent, age, and references. Appendix C is written by Felix P. Bentz and Judson B. Hughes, Santa Fe Minerals Inc. It interprets several reflection seismic sections in the delta region and gives further evidence to the presence of the impressive late Miocene canyon through which the earliest Nile, the Eonile, flowed and whose existence the author had previously postulated.

I-1. THE NILE BASIN

The Nile has a length of 6825 km from the sources of the Luvironza River in Tanzania to the shores of the Mediterranean (Figure 4). In this long course, the river follows generally a south to north path; both its source in Equatorial Africa and its mouth in the Mediterranean Sea lie within one degree on the same meridian of longitude. It crosses 35 degrees of latitude, drains an area of close to 3,000,000 km^2 and connects regions which differ from each other in relief, texture, climate, and geological structure. It derives its waters from the Lake Plateau which constitutes the southern swell bordering the Sudan basin and from the Ethiopian highlands which form part of the East African coalescing series of plateaus traversed by the great African rift system. In its northward journey, the Nile drains the major interior basin of the Sudan across the high Nubian swell into Egypt and the Mediterranean by way of a series of cataracts.

The Nile is a salient feature of the African continent which forms a vast continental shield margined with folded belts in its extreme north and south. Save for limited marine incursions along its coastal plains, Africa has remained a land area since the Ordovician and much of it since the Precambrian. The movements that affected the continent were epeirogenic, giving a structural pattern of broad basins separated by irregular swells (Figure 5). The plateaus and swells were intermittently uplifted and de-

Table I-1. Principal Rivers of the World

River	Length (km)	Basin (km^2)	Annual discharge (billion m^3/year)
Nile	6825	3,050,000	86
Amazon	6700	7,050,000	3000
Congo	4100	3,700,000	1400
Mississippi	3970	3,220,000	600
Zambezi	2700	1,300,000	500
Mekong	4200	795,000	400
Danube	2900	1,165,000	200
Niger	4100	1,890,000	180
Rhine	1320	162,000	80

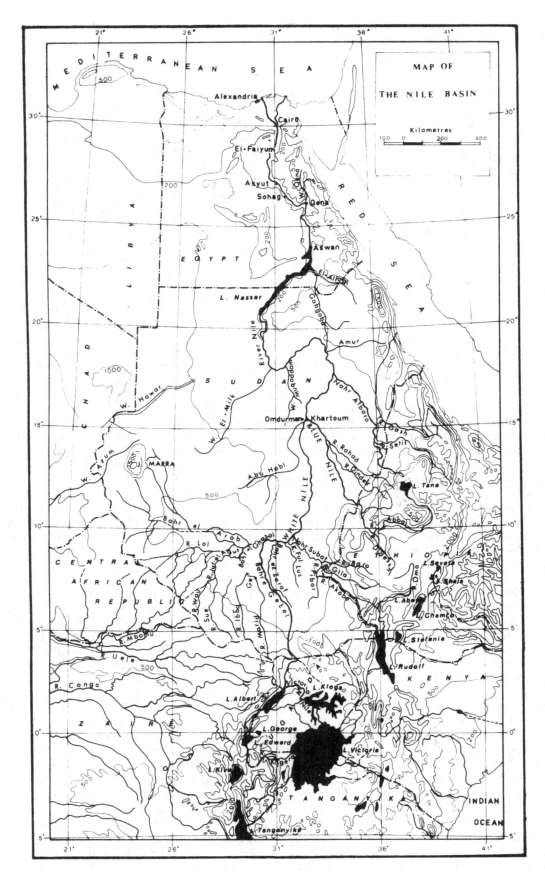

Figure 4. Map of the Nile basin.

Figure 5. Map showing in a generalized way the tectonic basins of Africa and intervening swells, plateaus, and rift valleys. After Holmes (1965).

nuded, with the result that they now consist largely of old rocks which were formerly deep seated. The basins were the receptacles of thick continental deposits representing the material eroded from the uplifted swells. Some of the African basins have internal drainage such as the Chad, El-Juf, and Kalahari-Kubango basins. Others are copiously watered and possess external drainage. The Congo basin is traversed by the great Congo River which cuts its course in the thick continental sediments of Karroo age and escapes to the Atlantic across the western swell by a series of cataracts.

The basin of the Sudan is less readily traversed by the Nile. In the southern part of the basin, the White Nile follows a sluggish and tortuous course through a wide expanse of swamps and lagoons (the Sudd). These swamps are the remains of a former lake, the extent of which is marked by vast spreads of sediment. Today the basin is drained by the trunk channel of the Nile which overflows through a notch in the northern rim of the basin to become the vigorous Nile of the six cataracts between Khartoum and Aswan.

The Nile, therefore, negotiates its way through five regions which differ from one another in structure and geological history. These are from south to north: the Lake plateau, the Sudd and Central Sudan, the Ethiopian highlands, the cataracts, and the Egyptian region.

The Lake plateau represents the northern part of the great African swell which extends from beyond the fault depression of the Limpopo Valley, Southern Rhodesia, across the Zambesi to the great Central plateau of Tanganyika, Kenya, and Uganda. This vast region was reduced to a pediplain in Miocene times, but then it was upwarped by about 2000 m and fractured by the African Valley system. Lake Victoria occupies a structural sag between the eastern and western swells which were affected by the deep rift valleys of Africa. Despite its enormous size, the Lake's greatest depth is no more than 90 m. For long periods the lake had no outlet and it is only since about 12,000 years B.P. that it has drained into the Nile (Kendall, 1969) through the Ripon Falls. The Nile then cuts across the many-armed shallow Lake Kyoga and then descends rapidly to Lake Albert via the Murchison Falls. Leaving Lake Albert, the river flows placidly for about 225 km between marshy banks to Nimule, and then proceeds by a series of rapids for about 210 km to reach the Sudan plains at Juba. The level of Lake Victoria is 1135 m above sea, and that of the river in flood at Juba 455 m. The total fall of the river in the 820 km of its course between these two places is 680 m. The geological history of this region is extremely complicated and was affected by Tertiary and recent movements of the great African rift. The drainage of this plateau into the Nile basin took place only after its uplift and after the formation of Lake Victoria. Previous to this uplift the plateau's drainage seems to have been toward the Congo. Rivers draining into Lakes Kyoga, Albert, and Victoria have reversed drainage.

The Sudd region extends from Juba to Malakal for a distance of 905 km. It is a region of swamps and marshes occupying the southern part of the great alluvial plain which stretches northward with a gentle fall from the Lake plateau region and spreads out laterally for more than 220 km on either side of the river. The annual rainfall over the Sudd region averages about 900 mm. The northern part of the Sudd region, which is called Bahr el-Gebel, is joined by two great tributaries, the Sobat and the Bahr el-Ghazal, the former bringing the discharge of rivers coming from the Ethiopian highlands, and the latter of rivers coming from the Nile-Congo and Nile-Chad divide. The flow of all these rivers is greatly hampered by widespread swamps of tall grasses and papyrus, causing a loss, by evaporation and transpiration, of a large proportion of the water which would otherwise reach the Nile by the Sobat and the Bahr el-Ghazal. Navigation both of the tributary rivers and of the main river is often rendered extremely difficult by detached masses of rotting vegetation. It is this circumstance that gives the Sudd region its name (the word "Sudd" in Arabic means blockage or stoppage). The level of the Nile in flood is 455 m at Juba and 386 m at Malakal. The total fall of the main river in its passage of 958 km through the Sudd region is 69 m, corresponding to an average slope of 1:13,900 which, it may be remarked, is flatter than the average slope of the river in its passage through the Egyptian region from Aswan to the Mediterranenan.

The Central Sudan region extends from Malakal northward to Khartoum, a distance of 809 km. It lies to the north of the great alluvial plain of the Sudd region. It differs from the latter in having a drier climate (average annual rainfall about 500 mm) and in being free from swamps, so that large areas can be cultivated by irrigation. In its course through the Central Sudan, where it is called the White Nile, the Nile receives no tributaries; but at Khartoum, on the northern limit of the region, it is joined from the east by the Blue Nile. This powerful tributary, which comes from the Ethiopian highlands, constitutes the principal source of the waters which cause the annual rise and fall in level of the river downstream of the junction. This is due to the seasonal fluctuations of the flow of the Blue Nile corresponding to seasonal fluctuations of pluviometry over the Ethiopian highlands. The level of the Nile is 386 m at Malakal and 378 m at Khartoum; the total fall of the river in its passage of 809 km through the Central Sudan region is only 8 m, corresponding to the phenomenally flat average slope of 1:100,000, or less than 1 cm/km. This has led many authors to speculate that a vast lake must have once covered this region as well as the Sudd region. The position and extent of this lake is discussed in length by Ball (1939).

The Cataract region extends from Khartoum to Aswan, a distance of 1847 km as measured along the curved course of the river, but only 950 km in a direct line. It is a hilly desert region with a very dry climate, the average annual rainfall being less than 50 mm. The Nile flows through this region in a well-defined valley in which stretches of cataracts alternate with more placid reaches. These cataracts, which occupy 565 km of the 1847 km of the river's course through the Cataract region, are due to outcrops of hard crystalline rocks which offer far greater resistance to the erosive action of the river than do the Nubia sandstones (Mesozoic sedimentary rocks) of the intervening stretches.

It is in one of the more placid reaches of its course through the Cataract region that the Nile receives its last tributary, the Atbara. This is a torrential seasonal stream having its sources in the Ethiopian highlands and joining the Nile about 322 km downstream of Khartoum. The level of the Nile in flood is 378 m at Khartoum and 91 m at Aswan (before the construction of the dams); the total fall of the river in its passage through the Cataract region is 287 m in 1847 km, or an average slope of about 1:6440. The action of the river in the cataract region, since at least the late Pleistocene, differs markedly from the regions north and south of it. In these latter regions the river is gradually raising its bed by deposition of silt, while in the Cataract region it is gradually lowering its bed by erosion. At the cataract of Semna (about 70 km south of Wadi Halfa) there is evidence, from inscriptions cut in the rock by Amenemhat III, that the Nile lowered its channel in the rocky barrier by about 8 m within the last 3800 years (Wilcocks, 1904; Ball, 1939).

Of the total course of the Nile, which has a length of more than 6800 km from its source near Lake Tanganyika to its mouth in the Mediterranean, only the terminal 1530 km lie within the borders of Egypt; and throughout this part of its course, the river receives not a single tributary. After entering Egypt from the Sudan at Wadi Halfa, the Nile flows for more than 350 km in a narrow valley bordered by abrupt cliffs of sandstone and granite before reaching the first cataract which commences about 7 km upstream of Aswan. Narrow strips of alluvial land, until recently, could be cultivated on either bank of the river in many parts of the Aswan-Wadi Halfa reach; but these, together with other features below the 180 m contour, are now completely drowned owing to the conversion of a long stretch of the valley upstream of the first cataract into a reservoir by the construction of the High Dam across the river at the head of the cataract. The cataract itself is a series of rapids resulting from the obstruction of the passage of the river by a multitude of rocky islands. Downstream of the cataract, the valley begins to broaden, and flat strips of cultivable land between the river and the cliffs gradually increase in width northward. Sixty km to the north of Aswan lies the Silsila gorge bound on both sides by quartizitic sandstone of the Nubia Sandstone which formed in antiquity one of the best quarries of this material. The Silsila seems to have formed a waterfall until the Holocene and, as a result of its damming effect, the tectonic graben of Kom Ombo, lying upstream, was flooded and formed a swamp of considerable extent in late Pleistocene time.

Near Esna, about 160 km downstream of Aswan, the sandstone of the bounding cliff gives place to limestone; and at Qena, about 120 km downstream of Esna, the river makes a great bend with limestone cliffs rising to heights of more than 300 m on either side. Past the Qena bend, the Nile swerves in the direction of the Red Sea and the valley broadens appreciably. It is significant that the river clings tenaciously to the right-hand side of the valley, sometimes washing the foothills of the mountainous Eastern Desert. Nearly 90% of the cultivable area in Upper Egypt thus stretches on the left bank along the margins of the western (Libyan) Desert. Near Assiut, some 260 km downstream of Qena, the cliffs on the western side become much lower than those on the east, and continue so for about 400 km to Cairo. The total area of Upper Egypt is barely 12,000 km², stretching over a distance of more than 1000 km.

Only 170 km long, the delta is twice the area of Upper Egypt. Beyond the apex it spreads in a plain studded with an intricate network of canals and drains; the former lie along the higher tongues of land, the latter in the hollows. Throughout recorded history the canal mesh has varied a great deal; many fossil branches are recognized. Today, the delta tributaries are the western Rosetta branch, about 239 km long, and the eastern Damietta, about 6 km longer. The whole mesh loses itself in a coastal marsh belt of waste land, the Berari, punctuated with a number of coastal and inland lagoons.

I-2. PREVIOUS LITERATURE

The literature on the geological history of the Nile is extensive. Among the more important workers who advanced theories to explain the origin of the river mention is made of Hull(1896), Judd (1897), Lyons (1906), Arldt (1911), Gregory (1920), Blanckenhorn (1901, 1910, 1921), and Awad (1928). More recent works include Huzayyin (1941), Sandford and Arkell (1929, 1933, 1939), Sandford (1934), de Heinzelin (1968), Butzer and Hansen (1968), Wendorf and Schild (1976), and Said (1975).

Blanckenhorn's work on the subject is one of the earliest attempts to formulate a theory of the development of the Nile based on observations in many areas in northern Egypt. According to Blanckenhorn, the River Nile occupies the

place of the marine Pliocene gulf which was formed as a result of the dislocations which took place in Egypt in early Pliocene time. Earlier drainage systems which preceded this post-Pliocene river were related to an old mighty river which Blackenhorn calls the "Libyan Ur Nil." Its course lay somewhere to the west of the present river. This early river is sometimes referred to in the writings of later workers as the Protonile, a term which in the present book will be used in a different sense and will be retained for a Quaternary river which filled the Nile Valley itself. According to Blankenhorn, the history of the Nile started with a gulf phase when a marine ingression filled the tectonic basin of the Nile up to lat. 29° N. At a later stage (lower Diluvium DI) the gulf was filled with the deposits of a locally fed river which was responsible for the deposition of the *Melanopsis* Stufe. The fauna of the "lower Diluvium *Melanopsis* Stufe" is devoid of the typical Central African forms.This led Blanckenhorn to advocate that the Nile during that time was not connected to the southern portion of its present basin, receiving its water and sediment from Egyptian and Nubian highlands. The deposits of the subsequent phases of the river are preserved in the form of three main groups of terraces which Blanckenhorn named high, middle, and lower gravel terraces. These are ascribed to the Diluvium (DII, DIII, and DIV, respectively). The silts constituting the modern alluvial lands of Egypt are classified by Blanckenhorn as "Alluvium."

Sandford and Arkell (1929–1939) studied the geology of the Nile Valley and the Faiyum divide. They, like Blanckenhorn, believe that the Nile cut its channel in the deposits of the Pliocene gulf which developed along the course of the Nile up to the latitude of Aswan. At the close of the Pliocene and up to the early Pleistocene epoch the Nile eroded its channel downward in the Pliocene sediments leaving behind terraces at heights of about 140, 115, 90, 60, and 45 m, respectively, above the present-day flood plain. Sandford and Arkell recognize in post-early Pleistocene time three groups of sediments that they ascribe to the early, middle, and late Paleolithic. The early Paleolithic sediments are recognized in the form of two gravel terraces, 30 m and 15 m higher than the flood plain, the latter of which contains Acheulian implements. The middle Paleolithic is made up of two gravel terraces, the higher of which can be followed at a uniform height of about 9 m above the present-day flood plain from Aswan to Assiut, north of which it appears to have been removed by denudation. The lower terrace, containing Mousterian implements, is traceable at a height of about 3 m above the present-day flood plain between Aswan and Luxor, but further north appears to descend below it. The upper Paleolithic is represented by terraces made up mainly of silts containing Sebilian implements. These silts are recognized from Wadi Halfa northward. They rise about 30 m above the present-day flood plain in Nubia and gradually diminish in height northward, reaching to about 6 m above the present-day flood plain at Luxor, and practically coinciding in level with it at Nag Hamadi. These silts were later dissected by the Nile to lower elevations in middle–upper Paleolithic time reaching depths below the present-day valley, to be followed by an aggradational episode which continued to the present.

Ball (1939) uses the data of Sandford and Arkell to reconstruct the history of the Nile and delta speculating as to the fluctuations of the sea levels and their impact on both the various Nile terraces and the extent of the delta. Ball estimates that by middle Mousterian time the Mediterranean sea level stood at −12 m relative to its present level. He further calculates that by Acheulian time the sea level stood at +25 m, by early Mousterian +18 m, by early Sebilian −43 m, and by early Neolithic −10 m relative to the present level.

Recent contributions to the history of the Nile Valley were made by Butzer (1959 a,b; 1960 b). More significant contributions were made by scientists who worked closely with the archeological excavations in the Kom Ombo area and in Nubia prior to its flooding by the Aswan high dam. Among the most important works are those of Said and Issawi (1964), de Heinzelin and Paepe (1965), de Heinzelin (1968), Butzer and Hansen (1968), Maley (1970), Said (1975), Smith (1976) and Wendorf and Schild (1976). De Heinzelin (1968) works out the fluvial deposits of the river in Nubia. He, like Said and Issawi (1964), recognizes that the fluvial sediments of the river fall into two major groups, an earlier sand–gravel complex which must have been deposited in a fluvial environment under a different regimen with a different source, and a later group of sediments which are similar to the sediments which accumulate in the present-day Nile. De Heinzelin names this earlier group of sediments the Dabarosa group of what he terms the Pre-Nile system. The term "Pre-Nile" in de Hein-

zelin's work is a collective term for all deposits preceding the deposits of modern aspect of the Nile. In the present work the term "Pre-Nile" is used for a group of sediments of one episode of the river's development which immediately preceded the modern Nile. The later Nile sediments are divided by de Heinzelin into several formations. The lowest formation is the Ikhtiariya dunes which cover the deeper parts of the eroded channel of the Nile about 13 m above the present flood plain. This is followed by three main aggradational deposits which are separated by recessional features. The oldest of these aggradational deposits is the Dibeira-Jer formation which occurs as benches lying between 27 and 35 m above the present flood plain. The type locality carries implements of Khormusan tradition (Marks 1968) and is dated 22,730 B.P. ± 280 years. The last stages of this aggradation are characterized by accumulation of calcified roots which de Heinzelin calls the Jer facies. The recession which followed brought the level of the river down to 11 m above the flood plain when the dunes of the Ballana formation started to accumulate. The second aggradation is termed the Sahaba, and it forms terraces made up of fluvial sands and gravel about 25 m above the flood plain. The last of the aggradations is the Arkin which started about 11,000 B.P. forming benches 13 m above the level of the flood plain and then dropping continuously and at a regular pace to the present-day level of the river. De Heinzelin also notes that the most recent of the sediments of the Nile (post-Arkin) possess a fauna which is different from that which thrives in the Nile today.

Butzer and Hansen (1968) study the geomorphology and sediments of the Nile Valley in Nubia and the Kom Ombo area. The Pliocene "Gulf Deposits" of Sandford and Arkell described from the west bank of the Kom Ombo area are taken by Butzer and Hansen to indicate a fluvial to lacustrine environment during the Pliocene. In addition to this Pliocene sandy facies Butzer and Hansen identify from the east bank a shale and evaporite facies which they consider to be of lagoonal origin (Kharit-Shait facies). Our observations in Kom Ombo and other areas, however, show that the sandy facies of the "Pliocene Gulf Deposits" exist in both the east and west banks of the river and that they are of fluvial origin representing, in fact, the deposits of the middle Pleistocene of Egypt (Q_2). On the other hand, Butzer and Hansen's Kharit-Shait facies is now considered to belong to the deposits of a yet earlier river which is dated as upper Pliocene (Paleonile) (Said, 1975). The fluvial nature of these deposits is proved in spite of the fact that they include gypsum and salt pockets. These pockets are of decided postdepositional aspect. Butzer and Hansen conceive the first alluvial deposits in the Nile Valley to be represented by the great gravel beds that are recorded on both the east and west banks of the Nile in the Kom Ombo area. Their presence on both the east and west banks is taken as an indication that the river had its sources in the Red Sea hills along the modern Wadi Shait-Wadi Natash drainage system. However, our observations show that the gravels on the west bank extensively recorded in the Darb el-Gallaba area are of different composition and age from those recorded on the east bank of the river. The polygenetic gravels of the eastern bank follow on top of the Prenile of late Pleistocene age (Q_2/Q_3), while those on the west bank are the deposits of an early Pleistocene Nile in Egypt (Q_1). Following upon the terrace gravels and red paleosols of the early and middle Pleistocene, true nilotic deposits mark the late Pleistocene and Holocene. These are divided by Butzer and Hansen into several aggradational alluvial deposits which are termed, respectively, the basal sands and marls (Korosko formation), the older flood plain silts (Masmas formation), and the younger channel silts (Gebel Silsila formation). These are followed by two units of wadi alluvia which are termed, respectively, the Ineiba and Shaturma formations. The Masmas and Gebel Silsila formations are proper nilotic deposits which correlate with de Heinzelin's lower aggradational episodes. The stratigraphic position of the Korosko formation, however, is not clear since its fluvial character is dubious. Butzer and Hansen suggest either a lacustrine or a mixed fluvial aquatic environment of deposition (p. 95) or a subaqueous environment along the valley margin (p. 96).

The Ineiba wadi alluvium is, in fact, a playa deposit which filled the blowouts in the Kom Ombo plain in the post-Gebel Silsila time. The Ineiba formation is certainly correlatable with the Dishna playa deposits of Said, *et al.* (1970).

Wendorf and Schild (1976) deal with the prehistory of the Nile Valley and work out a sequence of the later deposits of the Nile based on field work carried out in several areas in Upper Egypt and Faiyum, the preliminary results of which had been published earlier: Wendorf and Said (1967), Said *et al.* (1970, 1972a,b) and Wendorf *et al.* (1970a,b,c). According to the field work carried out by these authors, the aggradational silts recognized in Egypt had been correlated with those described in Nubia by de Heinzelin (1968). The lower aggradational silts had been correlated with the Dibeira-Jer and the upper silts with the Sahaba. These had been separated from each other by a recessional episode in which dune sand (the Ballana) and pond sediments and/or diatomites (Deir el-Fakhuri) are identified. The Sahaba is followed by the recessional playa deposits of the Dishna formation. Wendorf and Schild (1976) reconsider the sequence and decide that the lower aggradational silts in Egypt and the overlying Ballana dunes belong to one aggradational episode which they take to be coeval with the Masmas formation of Butzer and Hansen, and give it the name Masmas-Ballana and an age of 26,000 B.P. The Dibeira-Jer silts in Nubia are taken by Wendorf and Schild to represent an older aggradation which is not recognized in Egypt, although the Korosko formation of Butzer and Hansen may be coeval. Wendorf and Schild come to this conclusion because they consider that the Khormusan industry which is associated with the Dibeira-Jer is of middle Paleolithic rather than late Paleolithic age, and that the radiocarbon dates which had been obtained for this unit seem not to be reliable in the light of the earlier dates given by Irwin *et al.* (1968) for the silts carrying the Khormusan industry. Wendorf and Schild further lump the Deir el-Fakhuri formation (thought to represent the remains of a recessional episode separating the lower and upper silts) with the succeeding upper silt aggradation (the Sahaba). They correlate this new formation with Butzer and Hansen's Darau member of the Gebel Silsila formation and term it the Sahaba-Darau formation. In a recent publication Wendorf and Schild (1980) suggest a major revision of this classification. From evidence obtained from a study of the recent Nile section in Wadi Kubbaniya opposite Aswan, these authors suggest that the Masmas-Ballana and the Sahaba-Darau formations represent a single major aggradation contemporaneous with pronounced eolian activity.

Table I-2 gives the Nile Valley rock units described by the different authors and correlates them with the units recognized in this work.

Table I-2. Nile Valley Rock Units as Described in This Work Correlated with Units Described by Other Authors[a]

Age (B.P.)	Stage	Subdivision	Said (this volume)	Sandford and Arkell (1929–1939)	Butzer and Hansen (1968)	de Heinzelin (1968)	Wendorf and Schild (1976)
	Neonile	δ Neonile	Arkin			Arkin	Arkin
30		γ/δ recession	Dishna-Ineiba		Ineiba	Birbet	
		γ Neonile	Sahaba-Darau	12 and 21 m terraces	Gebel Silsila Darau	Sahaba	Sahaba-Darau
		β/γ recession	Deir el-Fakhuri				
		β Neonile	Masmas-Ballana	6 and 30 m terraces	Masmas	Dibeira-Jer (upper part)	Masmas-Ballana
		α/β recession	Korosko-Makhadma Gerza-Ikhtiariya			Ikhtiariya	
		α Neonile	Dandara	3 and 9 m terraces	Korosko		Korosko
130	Prenile/Neonile		Abbassia	15 and 30 m terraces	Gallaba (*pars*)		
200	Prenile		Qena	44 m terraces Pliocene gulf deposits (*pars*)			Qena Dandara
650	Protonile		Idfu	60, 90 and 115 m terraces	Gallaba Adindan and Dihmit terraces	Dabarosa	
	Paleonile/Protonile		Issawia Armant	Pliocene gulf deposits (*pars*)			
1800	Paleonile		Madamud/Kafr el-Sheikh/Gar el-Muluk		Kharit-Shait facies		
	Eonile/Paleonile		Kom el-Shelul	Pliocene gulf deposits (*pars*)			
5400	Eonile		Qawasim				

Left-hand margin Quaternary designations: Q₃, Q₂, Q₁, Tpl, Tmu.

[a] Figures on the left-hand column indicate age in thousand years B.P.

CHAPTER II

The Nile in Egypt

The following chapters deal with the geological history of the river which followed a course closely associated with the present-day Nile Valley. The valley seems to have been cut during late Miocene time. The earlier pre-late Miocene rivers which drained the elevated lands of Egypt do not seem to have been associated with the valley in its present form and, therefore, do not constitute the subject of this work. Relics of these earlier rivers are preserved in fluviatile sand and gravel spreads and in deltaic sediments recorded in several places in northern Egypt. The earliest of the fluviatile Tertiary deposits date back to late Eocene and early Oligocene time. They are in the form of a 310 to 470 m thick section of deltaic and associated offshore and barrier beach deposits (Qasr el-Sagha and Qatrani formations) in the Faiyum region to the north of Birket Qarun at an elevation of 350 m above sea level. The Faiyum deltaic deposits have been extensively studied (Beadnell, 1905; Vondra, 1974). The most interesting feature of these deposits is their inclusion of important and unique vertebrate remains that have attracted the attention of numerous scholars (Beadnell, 1901, 1905; Stromer Von Reichenbach, 1902, 1907; Andrews, 1906; Simons and Wood 1968). No fluviatile sediments have been recorded so far that would give a clue as to the course of the late Eocene-Oligocene river that made up this delta which must have formed along the edge of a marine gulf that protruded inside the elevated lands of Egypt up to the Faiyum region. Recent drilling in the area extending from the Mediter-ranean to the Faiyum gives the first indication of the presence of this embayment by the discovery of the only record so far known in Egypt of marine Oligocene. Blanckenhorn (1901) is the first to visualize the presence of this bay, and he and Beadnell (1905) attempt to reconstruct the course of the river that debouched into this estuary claiming that it probably took its headwaters from lakes in Bahariya oasis in which the iron ores known in that region must have been deposited. Recent studies, however, show that the ores are not lagoonal in origin (Said and Issawi, 1965) and that their age is earlier than that of the delta of the Faiyum.

River deposits in the form of extensive sand and gravel spreads that overlie the deposits of this early late Eocene-Oligocene river are known in the Western Desert of Egypt over the plateau which extends between Minia and Bahariya oasis. They form part of the "gravel spreads" marked on the new geological map of Egypt (as a patch on this plateau) trending in a northwestern direction (Said, 1971). These deposits form part of the fluviatile facies of the lower Miocene deposits of the Western Desert of Egypt (Said, 1962a,b). These fluviatile deposits represent rivers which probably opened into the delta at Moghra oasis at the eastern tip of the Qattara depression. Here again the presence of land and semiaquatic fossil vertebrates in Moghra shows that the deltaic deposits of this area were formed by a prograding delta.

The field mapping of the fluviatile and associated sediments of the Nile Valley and the examination of a large number of boreholes both deep and shallow show that it is possible to conceive of the Nile as having passed through five main episodes since the valley was cut down in late Miocene time. Each of these episodes was characterized by a master river system. Toward the end of each of the first four episodes (the last is still extant) the river seems to have declined or ceased entirely to flow into Egypt. These five rivers are here termed the Eonile (Tmu), Paleonile (Tplu), Protonile (Q_1), Prenile (Q_2), and Neonile (Q_3).

The first of the rivers, the Eonile, was a late Miocene feature which was responsible for the cutting of the modern valley to great depths which have been recently fathomed in some parts of the valley. The depth of the Eonile canyon in northern Egypt reaches about 2500 m (see Appendix C). No deposits of the Eonile system are known in outcrop since the river was degrading its bed to the new lowered base level of the desiccated Mediterranean.

The sediments belonging to the succeeding river, the Paleonile, consist of a long series of interbedded, red brown fluviatile to fluvio-marine clays and thin, fine-grained sand silt laminae which crop out along the banks of the valley and many of the wadis which drain into it.

The recognition of the deposits of the succeeding three rivers makes possible the division of the Pleistocene of the Nile into a threefold system. The deposits of each of these rivers are distinct in lithology, stratigraphic relationships, and mineral content. They are separated from one another by great unconformities and long periods of recession. The earliest of these rivers, the Protonile (Q_1), was a highly competent river which carried cobble and gravel-sized sediments made up mainly of quartz and quartzites. The deposits of the succeeding river, the Prenile (Q_2), are mainly made up of massive cross-bedded sands (Qena formation). The deposits of the last of the rivers, the Neonile (Q_3) which is still extant, are indistinguishable from those of the present-day river.

II-1. THE EONILE (Tmu)

The cutting of the valley of the Nile seems to have taken place during the Messinian (late Miocene) time. This age has long been recognized as an episode of regression and erosion in northern Africa. Barr and Walker (1973) review the earlier records of Messinian in northeastern Africa and come to the conclusion that all lack definitive paleonotological evidence. In the case of Egypt, in particular, these authors agree with the current accepted view of the complete absence of marine late Miocene deposits in Egypt. The Deep Sea Drilling Project Cruise Leg 13 covering the Mediterranean Sea revealed the widespread occurrence of an evaporitic suite beneath the bottom of the sea which proved to be of late Miocene age. The evaporitic suite obtained from drilling during Leg 13 comprises halite, gypsum, anhydrite, and dolomite. Hsu, *et al.* (1973) Hsu and Montadert (1977) Ryan (1976) and Ryan and Cita (1978) give evidence that these were deposited in a "desiccated deep-sea basin model" which allowed the formation of these evaporites in a series of shallow salt lakes or playas representing the relics of a desiccated Mediterranean. According to these authors, the evaporites were formed in deep (3000 m or more) dry basins more or less similar in depth and configuration to the present floor of the Mediterranean. Nesteroff (1973), however, believes that the evaporites were formed in shallow depressions, no more than a few hundred meters deep, whose present-day bottom topography resulted from major tectonic subsidence during the Plio-Pleistocene. Stanley (1977) indicates that the evaporites were formed at intermediate depths in settings that changed considerably in the post-Miocene in response to both vertical tectonics and lateral sea-floor displacement.

All available evidence indicates that the Mediterranean sea level was lowered during the late Miocene by several hundred meters. This lowering of the sea level has supporting geomorphological evidence from adjacent land areas where channels cut by streams rejuvenated during the regression of the sea are recorded from all areas draining into this emptied Mediterranean basin.

Channels cutting deep into the elevated north African plateau and graded to this new base level are recorded from several places in Libya and Egypt. Squyres and Bradley (1964) record an abandoned channel in the Qattara scarp. Bellini (1969) records another channel in the area between Jaghbub and Giallo in Libya, while Barr and Walker (1973) report a deeply incised channel south of the Cyrenaican platform to the northeastern flank of the Sirte Basin in Libya which has cut more than 430 m below sea level into middle Miocene rocks. These latter authors suggest a sudden drop of considerably more than 430 m in the Mediterranean sea level to explain this deep drainage erosion in late Miocene time. The same authors also report that geophysical crews prospecting for oil along the eastern margin of the Sirte basin, northern Libya, experienced severe velocity problems in the upper 1000 to 2000 feet of surface sediments indicating rapid lithologic changes. These sharp velocity changes coincide in many cases with dry drainage systems that seem to be of considerable magnitude and depth. Farther east, a river canyon filled with early Pliocene sediments near Gaza is recorded by Gvirtzman (1969) and another is recorded by the same author in northwest Sinai under dune sands (Gvirtzman and Buchbinder, 1977).

The drillholes in the delta and valley of the Nile indicate the great depth to which the Nile channel was excavated during late Miocene time. South of Aswan and in the abandoned channel between Shallal and Aswan the granitic bedrock was hit at a depth of 170 m below sea level (Said and Issawi, 1964; Chumakov, 1967). To the west of Cairo at Tamuh water well the middle Eocene bedrock was reached at a depth of 568 m (Said and Yousri, 1964). In Assiut seismic crews prospecting for oil reported loose sediments about 800 m thick overlying bedrock along the mouth of Wadi Assiuti, one of the major wadis draining into the Nile (Squyres, personal communication). To the north of Cairo reflection seismic data indicate the presence of a deep channel at a depth of 2500 m below sea level. This channel was traced for some distance to the north (see Appendix C for further details about this channel). It is certain that this channel formed the bed of the Eonile which followed a northward course to the Mediterranean.

Figure 6 is a structure contour map of the base of the late Miocene in the delta region. It shows that at the advent of this age the southern part of the delta (here termed the South Delta block) was a positive area which represented in fact the northern margin of the African continent. To the north and beyond the cliffy edge of this block a deep embayment (here termed the North Delta embayment) covered the northern reaches of the delta and formed part of the Eastern Mediterranean basin. The southern limit of this basin seems to have been controlled by a complex series of west–northwest faults forming a hinge zone. Recent drilling in the North Delta embayment revealed the presence of an evaporitic suite (the Rosetta evaporites) at ex-

Table II-1.

No.[a]	Well	Depth (m)
24	Abadiya #1	3040
10	Abu Qir #IX	2486–2530
12	Baltim #1	3495
4	Busseili #1	2200–2500
1	el-Tabia #IX	2034–2059
13	Kafr el-Sheikh #1	3895–3963
25	Matariya #1	2290–2304
21	Qawasim #1	2800
8	Rosetta #1	2678–2720

[a] Number of well as it appears in Table A-1 and in all figures.

actly the same horizon recorded beneath the Mediterranean Sea by the Deep Sea Drilling Project Cruise Leg 13 at depths ranging from 3963 m (Kafr el-Sheikh #1) to 2059 m (el-Tabia IX) below sea level. Table II-1 gives the depth of the Rosetta evaporites in the delta wells.

The Eonile must have eroded its bed deep into the elevated Egyptian plateau along a path which defined the course of the present-day Nile. It must have also received waters from the numerous wadis which debouch into the valley today; their courses and depths must have been determined during that age. Many of these wadis are filled with sediments of the succeeding rivers which abut against their slopes indicating that these valleys were already in existence when these rivers started to flow. The Eonile seems to have cascaded over the northern cliffs of the South Delta block into the gradually desiccating late Miocene Mediterranean. It flowed in a canyon of great dimensions which extended from Aswan to the northern reaches of the South Delta block. At the end of the late Miocene the waterfall traversing the hinge zone retreated upstream and graded its bed. Figure 8 is a structure contour map of the top of the late Miocene (base Pliocene). It shows the path of the assumed canyon of the Nile in the delta region, its depth and width. It is interesting to note that its path coincides with the seismic belt defined by Gergawi and Khashab (1968a). This gives credence to the view that the path of the Eonile was originally deter-

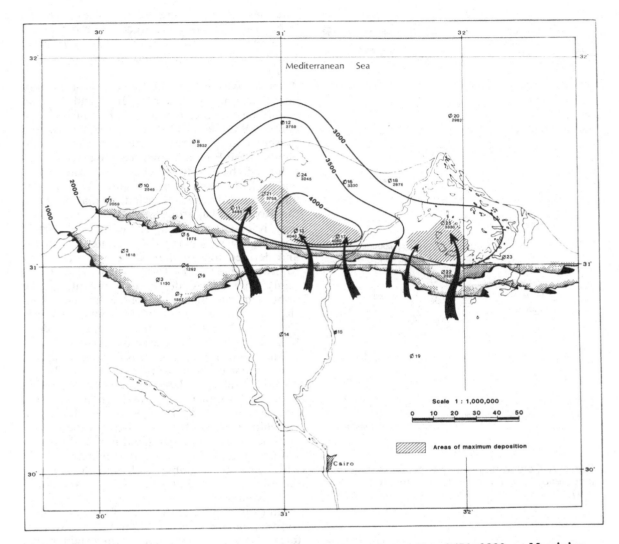

Figure 6. Structure contour map of base of late Miocene (Tmu). Numbers refer to wells, the locations and stratigraphic data of which are given in Table A-1.

mined by tectonics and was then eroded by the river.

The sediments of the late Miocene Eonile are in the form of coarse clastics which are known in the North Delta embayment in the subsurface. These are followed by evaporites (the Rosetta evaporites). The clastic sediments form the Qawasim formation (Figure 7). The Qawasim in its type locality (Qawasim well 1, lat. 31°20'07"N; long 30°50'55" E) has a thickness of 965 m (from depth 2800 to 3765 m). The following is a description of the Qawasim and the overlying Rosetta evaporites in Qawasim well 1.

Rosetta Evaporites: 2651–2800 m–Messinian

The Rosetta evaporites are made up of a lower 7-m thick gypsum bed which is overlain by a quartzose sandstone unit with interbeds of shale. It underlies unconformably the Pliocene *Sphaeroidinellopsis*-bearing beds. The unit is nonfossiliferous and is separated by virtue of the fact that it represents a fluvial to Sabkha environment of deposition different from that of the overlying and underlying beds. It is correlated with the Rosetta evaporites.

Qawasim Formation Unit I: 2800–3521 m–Messinian

This unit is made up of alternating beds of coarse-grained quartzose, poorly sorted sandstone with pebbles and derived Cretaceous and Eocene fossils, and dark gray shale. The sand-

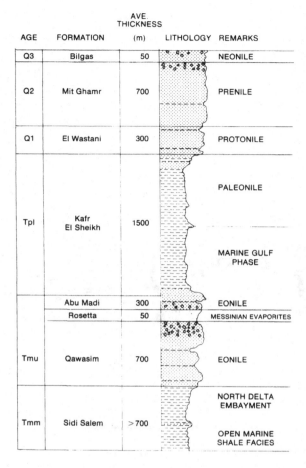

AGE	FORMATION	AVE. THICKNESS (m)	LITHOLOGY	REMARKS
Q3	Bilqas	50		NEONILE
Q2	Mit Ghamr	700		PRENILE
Q1	El Wastani	300		PROTONILE
Tpl	Kafr El Sheikh	1500		PALEONILE / MARINE GULF PHASE
	Abu Madi	300		EONILE
	Rosetta	50		MESSINIAN EVAPORITES
Tmu	Qawasim	700		EONILE
Tmm	Sidi Salem	>700		NORTH DELTA EMBAYMENT / OPEN MARINE SHALE FACIES

Figure 7. Composite columnar section of the subsurface deposits of the Nile delta. After Rizzini *et al.* (1978).

shale ratio is 6:5. Fossils are scarce and are localized in four levels: 2970–2990 m, 3100–3120 m, 3240–3282 m, and 3325–3365 m. The foraminifera separated from these levels are benthonic species of brackish water habitat which recur in all four levels: *Bolivina dilatata, Ammonia beccarrii, Uvigerina tenuistriata, Siphogeneroides* sp., and *Bulimina pupoides.* At levels 3100 and 3240 m some marine forms are observed: *Globigerina* and *Globigerinoides* spp.

Qawasim Formation Unit II :3521–3765 m–Tortonian

This unit is made up of alternating beds of dark gray to brown shales and poorly sorted quartzose sandstone. The sand:shale ratio is 3:5.

The fossils are few and are not diagnostic, but in the lower levels some stenohaline benthonic forms are observed: *Cibicides* sp., *Discorbis* sp., *Nonion* sp., and abundant *Ammonia beccarrii.*

The Qawasim formation has its type locality in this well between depths 2800 and 3740 m according to Rizzini *et al.* (1978) and between depths 2800 and 3765 m according to the present author. It represents the deltaic sediments of the Eonile. It overlies the marine sediments of the middle Miocene Sidi Salem formation and underlies the Rosetta evaporites or unconformably the Abu Madi formation when the Rosetta evaporites are absent. The lithology is clastic, thick layers of poorly sorted, coarse-grained sands with abundant pebbles and derived fossils alternating with dark gray to brown shales. The formation was most certainly affected by deltaic sedimentation; the upper part was most probably deposited subaerially at the foot slopes of the elevated South Delta block which stood up along the delta hinge line during the Messinian crisis.

The age of the formation is difficult to ascertain because of the lack of diagnostic fossils, but it is most likely of late Miocene age covering both the Tortonian and Messinian stages. This age assignment is inferred from the stratigraphic position of the formation which overlies unconformably well-dated middle Miocene sediments and unbderlies the evaporite beds which are correlatable with similar deposits recorded in the Mediterranean basin and dated as Messinian.

An examination of the logs of the delta wells shows a pattern of distribution of the Qawasim formation which sheds light on the origin of this formation (Figure 6). The lithology is uniform, coarse clastics alternating with fine-grained shales which must have been deposited by a river system during the late Miocene. The upper part of the Qawasim includes invariably more sand beds than the lower part. The Eonile must have cascaded over the cliffs and slopes of the hinge zone fully loaded to open up into the dry North Delta embayment distributing its load over its surface. As it flowed into the embayment where the slope seems to have been gentler, the river aggraded its bed thus prolonging it. The isopachous distribution of the late Miocene sediments (Figure 6) shows a pattern of a series of coalescing alluvial fans. The thickest part of each fan seems to have been at the footslopes of the hinge zone (Kafr el-

Sheikh, 1313 m; Bilqas, 1336 m; Sidi Salem, 1275 m; Qawasim, 1114 m; Matariya, 1040 m), becoming thinner toward the north. Along the areas which lie in the hinge zone the sediments are relatively thin (Qantara, 420 m; San el-Hagar, 513 m; Mahmudiya, 250 m; Busseili, 311 m). The front of the fan seems to have had an outline which was governed by the irregularities of the surface and by the interference of adjacent fans with one another at their confluent margins.

The coalescing alluvial fans of the Eonile were formed in front of the master stream as it emerged from the South Delta block. Other channels seem to have existed, and there is evidence from geophysical data that there were lesser tributaries pouring into the embayment at Matariya and Bilqas areas. It is interesting to note that the modern branches of the delta seem not to have been in existence in late Miocene time. In fact, the great canyon of the Eonile was to the east of the modern Nile in the southern part of the delta.

Table II-2 gives the thicknesses of the late Miocene deposits in the delta wells. A plot of these thicknesses indicates that the Eonile sediments accumulated in the overdeepened part of the North Delta embayment at the footslopes of the hinge zone.

II-2. THE EONILE/PALEONILE INTERVAL (Tmu/Tplu)

During the early Pliocene the rise in sea level, following the Messinian crisis, caused a marine ingression into the excavated canyon of the Eonile. There is evidence that this ingression extended inland in the form of an elongate gulf up to the latitude of Aswan. In the north, with the rise in sea level, the waters overflowed the banks of the canyon and covered large tracts of the areas surrounding the delta. Figure 7 shows the extent of this marine ingression in northern Egypt.

The early Pliocene sediments in the North Delta embayment are thick sand–shale deposits which carry a rich open marine fauna (Fig. 8). They follow on top of the late Miocene or older sediments with a marked unconformity which can be traced on all the seismic records with remarkable ease. The sediments include more sand members in their lower part (the Abu Madi formation) and are made up of shales in their upper part. This shale section forms the

Table II-2.

No.[a]	Well	Top Tmu (m)	Bottom Tmu (m)	Thickness (m)
24	Abadiya #1	2787	3295	508
16	Abu Madi #1	3025	3314	289
10	Abu Qir #IX	2464	2545	81
12	Baltim #1	3280	3759	479
17	Bilqas #1	2751	4087	1336
4	Busseili #1	2200	2511	311
1	el-Tabia #IX	1936	2059	123
18	el-Wastani #1	2752	2875	123
13	Kafr el-Sheikh #1	2735	4048	1313
5	Mahmudiya #IX	1725	1975	70
25	Matariya #1	2290	3330	1040
23	Qantara #1	1730	2150	420
21	Qawasim #1	2651	3765	1114
20	Ras el-Barr #1	2722	2982	250
8	Rosetta #1	2678	2832	154
22	San el-Hagar #1	1876	2389	513
11	Sidi Salem #1	2410	3685	1275

[a] Number of well as it appears in Table A-1 and in all figures.

lower part of the Kafr el-Sheikh formation. The early Pliocene sediments in the North Delta embayment range in thickness from 1388 m (Abu Qir well #IX) to 765 m (Bilqas well #1). Table II-3 gives the thicknesses of the early Pliocene in the different wells of the delta.

The foraminifera separated from these early Pliocene sediments in the embayment are described by Viotti and Mansour (1969) and Mansour et al. (1969). Three zones are recognized: the *Globigerina nilotica*, the *G. nepenthes,* and the *Sphaeroidinellopsis seminula grimsdalei.* In many wells the early Pliocene rests directly over the middle Miocene.

The early Pliocene sediments in the South Delta block vary in thickness and lithological composition from place to place depending on their position relative to the Eonile canyon. In Mit Ghamr and Shebin el-Kom wells, which lie

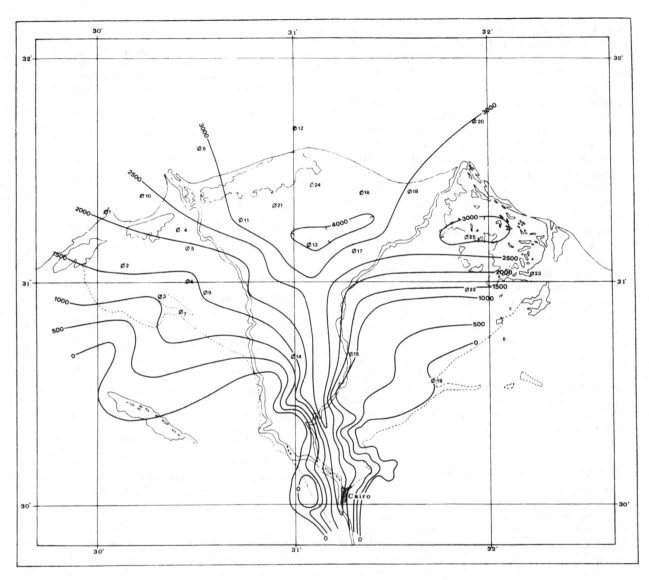

Figure 8. Structure contour map of base of early Pliocene (Tpll). Numbers refer to wells, the locations and stratigraphic data of which are given in Table A-1.

away from the Eonile canyon, the early Pliocene sediments are thin and attain thicknesses of 351 and 307 m, respectively. In the Abu Sir water well, situated at the western margin of the Eonile canyon near Cairo, a 98-m thick clay section (depths 475–573 m) occurs carrying a marine fauna ascribed to the early Pliocene (Said and Yousri, 1964). To the northwest of Cairo at the Abu Roash well #1 a 67-m thick

shale section (depths 94–161 m) carrying a marine early Pliocene fauna is recorded. The presence of this occurrence indicates that the sea drowned an old channel which the Eonile had formed as it negotiated its way to the north around the Cretaceous Abu Roash outlier which stood as a high in its path.

In the Nile gorge itself records of the marine Pliocene are given by Chumakov (1967) from as far south as Aswan. Chumakov describes from the deep channel, which he documents in the Nile Valley near Aswan; a 90-m thick section (depths 170–260 m) made up of basal sediments consisting of gray montmorillonitic clay with thin lenses of fine-grained micaceous sand

Table II-3.

No.[a]	Well	Top Tpll (m)	Bottom Tpll (m)	Thickness (m)
24	Abadiya #1	1723	2787	1064
16	Abu Madi #1	2214	3025	811
10	Abu Qir #IX	1076	2464	1388
12	Baltim #1	2112	3280	1168
17	Bilqas #1	1986	2751	765
4	Busseili #1	1440	2200	760
6	Damanhour, S. #IX	629	1222	593
7	Dilingat, N. #IX	495	1303	808
1	el-Tabia #IX	643	1936	1293
18	el-Wastani #1	2200	2752	552
3	Hosh Isa #IX	316	1069	753
9	Itay el-Barud #1	829	1519	690
2	Kafr el-Dawar #IX	308	1532	1224
13	Kafr el-Sheikh #1	1880	2735	855
5	Mahmudiya #IX	1229	1725	496
25	Matariya #1	1651?	2290	639?
15	Mit Ghamr #1	663	1014	351
23	Qantara #1	1450	1730	280
21	Qawasim #1	1760	2651	891
20	Ras el-Barr #1	2115	2722	607
8	Rosetta #1	1560	2678	1118
22	San el-Hagar #1	1100	1876	776
14	Shebin el-Kom #1	635	942	307
11	Sidi Salem #1	2000	2410	410

[a] Number of well as it appears in Table A-1 and in all figures.

and sandy loam rich in plant detritus. A uniform suite of authigenic minerals is present including glauconite, zeolite, pyrite, and siderite. The sands are partially cemented by secondary calcite. Rare ostracods belonging to the genera *Cypridea, Cyprinotus, Limnocythere, Eucypris,* and *Candoniella* are present, which suggests a brackish environment of deposition and an

early Pliocene age for the lowermost channel fill. Chumakov believes this channel to have been cut during late Miocene time, resulting in the formation of an enormous marine estuary in the Nile Valley during the early Pliocene. The presence of this marine estuary had long been inferred from the study of the distribution of the sediments of the Pliocene in Egypt (Blanckenhorn, 1921; Sandford and Arkell, 1939).

Along the delta edges there are limited records of the early Pliocene. In the Burg el-Arab well #1, 45 km west of Alexandria, the early Pliocene is represented by a 50-m thick shale section rich in marine faunas. It rests under the Pleistocene oolitic limestones and unconformably over the middle Miocene limestones (Omara and Ouda, 1969). In Wadi Natrun well #1, halfway along the Cairo–Alexandria desert road, the marine early Pliocene rests over sands that follow a basalt sheet of supposed Oligocene age.

Along the peripheries of the narrow gulf of the Nile shallow marine sediments, mainly marls, sandstones, and coquinal limestones occur. They make up numerous outcrops that skirt the cultivation and abut against the bounding Eocene rocks of the valley in its northern reaches up to the latitude of Beni Suef. The classical locality of these deposits is Kom el-Shelul to the south of the Gizeh Pyramids (Figure 9). The remarkable abundance of fossils in this locality makes it one of the most visited areas. It is described by Blanckenhorn (1921), Mayer-Eymar (1898), Sandford and Arkell (1939), and many others. The section, which forms the type locality of the Kom el-Shelul formation, begins by a basal 10-m oyster bed made up almost exclusively of *Ostrea cucullata* shells. This is followed by a 2-m sandstone bed crowded with *Pecten benedictus* and *Chlamys scabrella.* Upon this lies a sandstone bed only ¹/₂-m thick full of remains of *Clypeaster aegyptiacus,* casts of large gastropods (*Strombus coronatus, Xenophora infundibulum,* etc.), the same pectens in the underlying bed, and many other fossils. Then follows a 10-m bed of nonfossiliferous yellow quartzose sand and brownish sandstone with large flint pebbles.

Several other exposures are known along the cultivation edge on both banks of the Nile. Many of the exposures listed by Blanckenhorn (1901), Sandford and Arkell (1939), and Little (1935) have been tilled up or have become the sites of habitation of local villagers and are, therefore, difficult to reexamine. Perhaps the

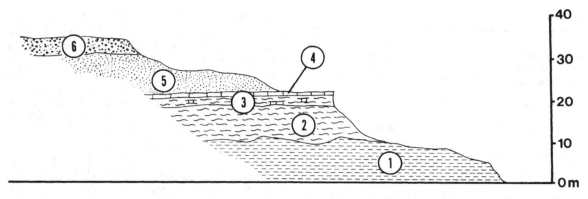

Figure 9. Section in the side of wadi at Kom el-Shelul, Pyramids plateau, Gizeh.
Q₁

6. Gravel terrace

Tpll

5. Sand, loosely cemented
4. *Clypeaster-Strombus bed*
3. *Pecten benedictus* bed
2. *Ostrea cucullata* bed

Teu

1. Late Eocene *Carolia* and *Ostrea clot-beyi* beds

best preserved of these exposures is in Helwan (Tebin) and the Nile-Faiyum divide. In Helwan the exposure is about 20 m thick and is made up of a series of marl, sandstone, shale, and thin limestone bands rich in fossils commonly associated with the Pliocene deposits of Egypt. Farag and Ismail (1959) describe the section and Said (1955) gives a list of the foraminifera separated from a marl bed in this locality.

Blanckenhorn (1901) and Little (1935) give several sections in Wadis el-Hai and Nuumiya on the east bank of the river opposite el Ayat and Iskar villages. Both sections are in the form of a thin (1 m) bed of sandstone or marl rich in *Ostrea cucullata* and/or *Pecten benedictus* which are topped by coarse conglomerates of local derivation. Both sections abut against middle Eocene limestone. In Wadi Sannur a 6-m section is described from this locality by Blanckenhorn. It is made up of sandstone and oyster banks as well as hardened shaly bands topped by a ¹/₂-m thick bed of boulders of local derivation. In Gebel Um Raqaba, the most southerly extension of the marine Pliocene exposures along the east bank of the Nile, a section of sandstones with oyster casts about 7 m thick is described. It overlies unconformably the upper Eocene and underlies an 8-m thick section of conglomerate and sandy marl.

On the opposite (western) bank of the Nile in the Nile-Faiyum divide several sections of Pliocene deposits are described by Sandford and Arkell (1929) and Little (1935). The best exposure is that of Shakluf bridge in the Gebel Naalun area (Figure 10). Here the section is about 36 m thick of which the top layers, made up of cross-bedded nonfossiliferous yellow to brown quartz sands (12 m) and a hard conglomerate bed of course gravel cemented by travertine (2 m), may not belong to the Pliocene. The lower part of the section is made up of thin sandstone beds, somewhat cemented sand beds alternating with thick blue to purple gypseous shales. The hard yellow sandstone beds include casts of *Pecten benedictus, Chlamys scabrella,* and *Ostrea cucullata* shells.

One of the most significant outcrops of the late Pliocene is that of Wadi Natrun where the early Pliocene sediments interfinger and are followed by the deposits of the succeeding river. In this locality the lower beds of Gar el-Muluk section carry the usual marine Pliocene fauna of the Kom el-Shelul formation. The elevation of this shallow marine Pliocene exposure is in the range of 4 to 6 m above sea level. In this respect, it assumes a very low altitude as compared to those outcropping in the Cairo-Faiyum reach on the western bank of the Nile and to those on the eastern bank. Thus, it seems to have been overlapped by the upper Pliocene river, the Paleonile, where it formed the site of a delta of that river.

Origin of the Nile Gulf

Although the origin of this elongate channel that was flooded by the early Pliocene ingression was the result of erosion by a river that had graded its bed to a lowered base level, it is certain that its path had been determined by tectonics. It has already been pointed out that the path of the Eonile in the delta reach lay along the mid-delta seismic belt delimited by Gergawi and Khashab (Figures 8 and 16). The Nile Valley itself lies along a seismo-active belt according to Gorshkov (1963) who lists the major historic earthquakes known to have taken place along the Nile Valley using the data

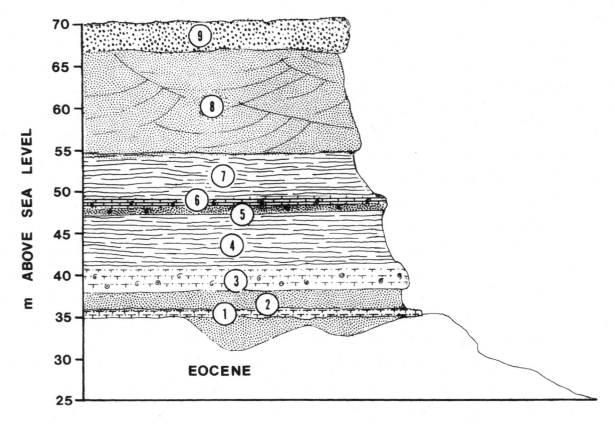

Figure 10. Section at Shakluf bridge, the western edge of the desert divide between Nile valley and Faiyum divide. After Sandford and Arkell (1929).

Q₂

9. Hard conglomerate made up of coarse gravel, cemented by travertine
8. False-bedded yellow, brown and white quartz sands and sandstones

Tpll

7. Blue and purple gypseous shaly clay, weathering mottled red and brown
6. Hard ferruginous limestone with *Cardium subsociale* and *Andara diluvii*
5. Brittle porous sand rock, rich in *Cardium linnei, C. echinatum, Meretrix chione, Venus plicata,* etc.
4. Blue and purple clay, same as bed 7
3. Two beds of hard yellow sandrock, enclosing a seam of soft yellow sand, the whole crowded with *Pecten benedictus, Chlamys scabrella,* and *Ostrea cucullata*
2. Yellow sand bed, locally cemented
1. Sandstone carrying a large number of casts of small bivalves, resting unconformably over a slipped Eocene block

published by the National Astronomical Center, Cairo (Ismail, 1960; Sieberg, 1932).

The presence of slipped masses and thick breccias at the foot slopes of the cliffs of the valley shows that these must have slid along the slopes by the trigger action of earthquakes. The slipped masses (called kernbuts by Lawson, 1927) reach large dimensions and are common along the valley. Their presence has attracted the attention of many workers. Sandford and Arkell (1933) give a description of many of these masses in the Thebaids of which the mass of Babein, which rises from the west bank of the Nile above Armant, is the most impressive. It is a detached mass from Gebel Rakhamiya which lies today on the eastern bank of the river. Little (1935) maps the blocks occurring along the shore of the northern part of the Pliocene Gulf and notes that the largest continuous relic of these masses runs for 7 km; at one place it is 1¹/₂ km wide. He further notes that most of these masses consist of upper Eocene rocks, but some are middle Eocene; and that "in a few places the beds are almost vertical and in others nearly horizontal, but the dip is generally between 15° and 5° toward the old cliff from which they came." He goes on to state that "since the landslides occurred the horizons from which they were derived have been carried back by erosion to the site of the present cliffs—distances varying from 10 to 20 km."

Similarly, the numerous reentrants along the cliffs which bound the Nile Valley in its middle course cannot be explained by normal weather-

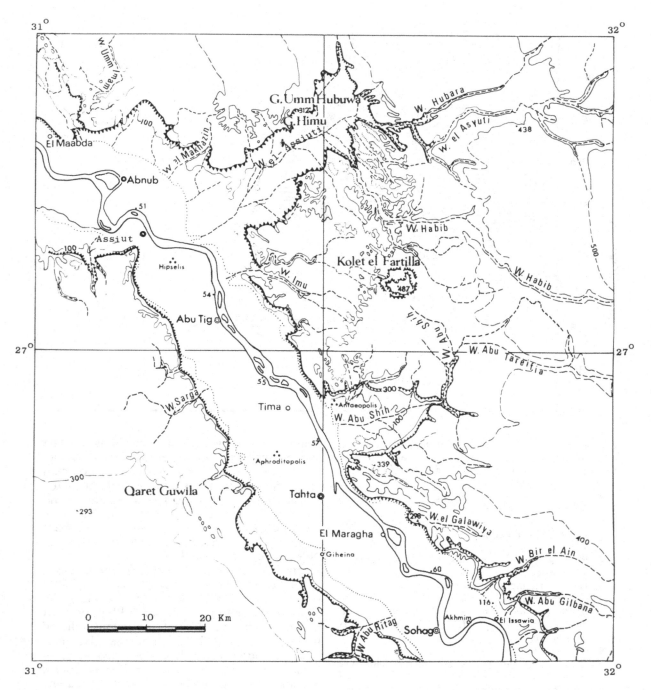

Figure 11. The Nile between Manfalut and Sohag showing the Tahta sphene and the Wadi el-Assiuti rhomb.

ing processes but must be attributed to seismic activity. These reentrants assume the shape of a rhomb (Nag Hamadi, Akhmim, Tell el-Amarna, Assiut) or a sphene (Tahta, Sannur). Many have no drainage into them but some form the mouths of some of the wadis which drain into the Nile. The fact that these reentrants (Figures 11 and 12) become suddenly wider as these wadis debouch into them shows that they cannot owe their origin to normal erosion. They are best explained by wrench faulting along an originally deflected or sinuous fracture which must have brought about a separation of the adjoining blocks and the development of gradually

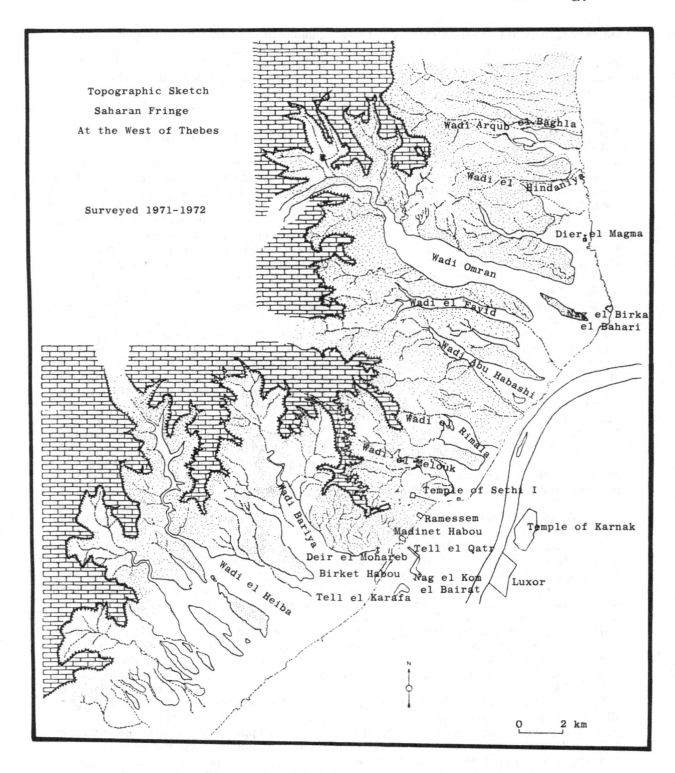

Figure 12. The cliff at Thebes (Luxor) showing distribution of Armant Formation (dotted). After Coque and Said (1972).

lengthening gaps (Quennell, 1958). According to Youssef (1968), horizontal displacements along the wrench faults were responsible for the formation of the valley of the Nile before the Pliocene. Figure 13 shows the centers of earthquake activity in Egypt which coincide to a remarkable degree with the areas that were affected by the submergence of the Pliocene gulf.

The structure of northern Egypt evolved during Oligocene and early Miocene times and has remained since then almost inactive except for minor tremors and earthquakes along some of the older lines. In spite of the fact that the Nile passes through a seismo-active zone no evidence of volcanic activity or major faulting is known from northern Egypt later than the early Miocene. Meneisy and Kreuzer (1974), who determined the radiometric age of a large number of basalts from northern Egypt, report no basalts younger than the early Miocene. Figure 14 gives schematically the major tectonic elements of northern Egypt. It shows that the northern part of the Nile delta (North Delta embayment) belongs to a separate unit forming part of the Levant basin which was affected by the large scale tectonics of the Eastern Mediterranean and the history of the closing up of the Tethys (Kenyon *et al.,* 1975; Neev, 1975; 1976; Neev and Friedman, 1978; Le Pichon *et al.,* 1973; Ross and Uchupi, 1977). This embayment, like the Levant basin, is characterized by an oceanic or semioceanic type of crust (Gergawi and Khashab, 1968b; Woodside and Bowin, 1970).

In contrast, the southern part of the delta belongs to the stable African platform which was elevated during the Oligocene and late Miocene to great heights along east–west faults deflecting toward the west of the embayment to a west–northwest direction. Toward the east of the embayment these faults merge into the seismo-active Pelusium line. The hinge zone, which separates the North Delta embayment from the South Delta block and the table lands of northern Egypt, is made up of a series of step faults. Great basalt sheets issue along many of these faults.

The distribution, isopachs, and disposition of all strata of post-early Miocene time (and even earlier times) were determined by this structure which influenced the advancing seas and the drainage lines of Egypt. Druing Oligocene and early Miocene times the sea advanced as far as the southern end of the evolving hinge zone causing the old drainage channels to become shallow and alluviating. It was only during the Messinian crisis that the Nile eroded its deep channel in the table lands of Egypt and in the South Delta block, cascading over the hinge zone by way of a series of water falls, and finally dumping its sediments in the North Delta embayment.

II-3. THE PALEONILE (Tplu)

Sediments belonging to the Paleonile River system consist of a long series of interbedded red-brown clays and thin fine-grained sand and silt laminae which crop out along the banks of the valley and many of the wadis which drain into it. These clays and silts are noted by previous workers and referred to in the literature as the *Melanopsis* Stufe by Blanckenhorn (1901, 1921), the chocolate-brown clays of the Gulf phase by Sandford and Arkell (1933, 1939), the plastic clay layer by Fourtau (1915) and Said (1973), the Kharit-Shait facies of the Pliocene gulf deposits by Butzer and Hansen (1968), or the Helwan formation by the Geological Survey of Egypt (Said, 1971).

Similar silts and clays are also reported from the subsurface in practically all the boreholes drilled in the valley and delta. They occur consistently below the sand–gravel layer which underlies the agriculural silt layer characteristic of the fertile lands of Eypt. In the boreholes of the North Delta embayment these silts and clays pass imperceptibly into the silts and clays of early Pliocene gulf phase forming one rock unit overlying the Abu Madi formation which is different lithologically. The entire clay silt section is named the Kafr el-Sheikh formation. In its type section (Kafr el-Sheikh well #1, lat. 31°10′23″N; long. 31°04′55″E) the formation has a thickness of 1760 m (between depths 975 and 2735 m below sea level). The lower part of this unit is of early Pliocene age and includes a rich foraminiferal assemblage which belongs to the *Globorotalia margritae* and *G. punticulata* faunules. The upper 905 m of this unit include, in addition to the marine faunas, a brackish water foraminiferal assemblage which increases relative to the marine assemblage both in the number of species and of individuals as one progresses toward the upper layers of the section. This upper part was deposited under estuarine conditions by the Paleonile.

The paleontology of the Paleonile sediments suggests a gradual filling of the valley and delta by a river system that debouched into the marine Pliocene gulf with the fluviatile sediments progressively moving northward. In the south the formation carries a fauna of estuarine to

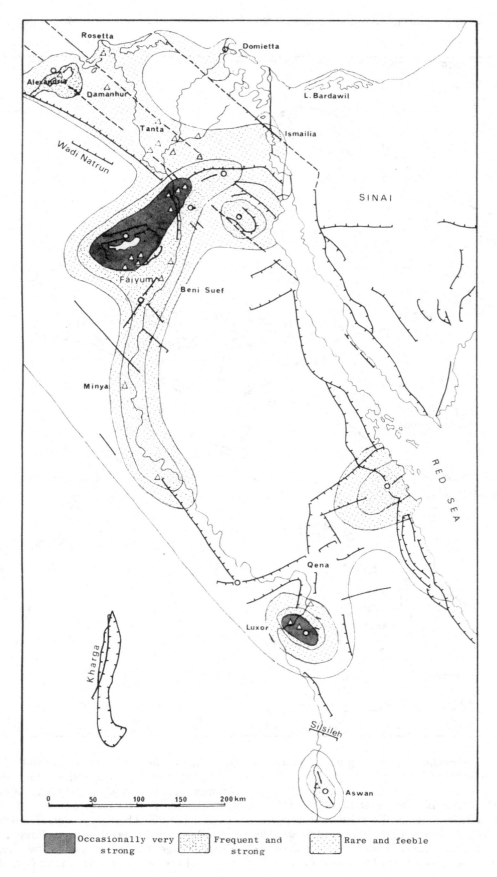

Figure 13. Seismicity map of Egypt. After Sieberg (1932).

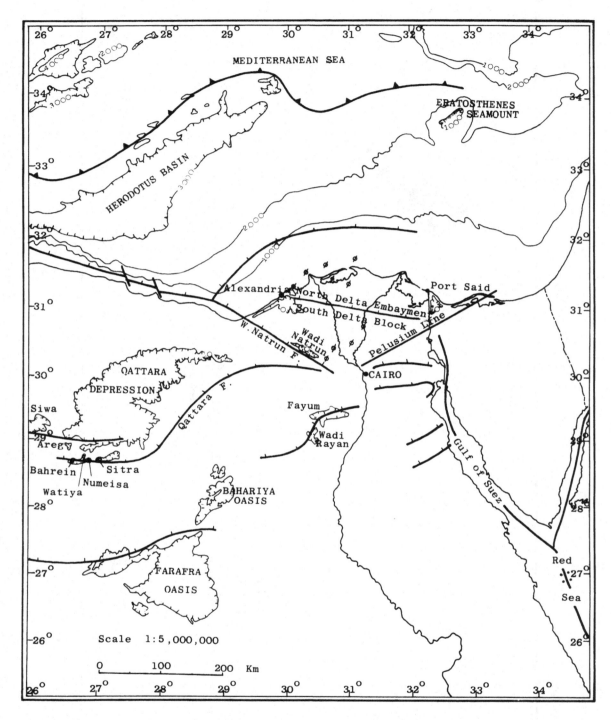

Figure 14. Schematic map showing major tectonic elements of northern Egypt.

brackish water habitat. Blanckenhorn (1901, 1921) describes some of the more characteristic fossils found in this unit: *Melanopsis aegyptiacus, Melania tuberculata, Hydrobia stagnalis, Vivipara martensi, Mactra subtruncata,* and others. Said and Yousri (1964) describe from a well near Cairo a rich brackish-water foraminiferal assemblage of boreal habitat and Atlantic affinities in the upper levels of the clays of the Paleonile. This can be correlated with the Castellarquato faunas now considered as of late Pliocene age (see discussions by Berggren, 1969; Hays *et al.,*1969; Ciaranfi and Cita,

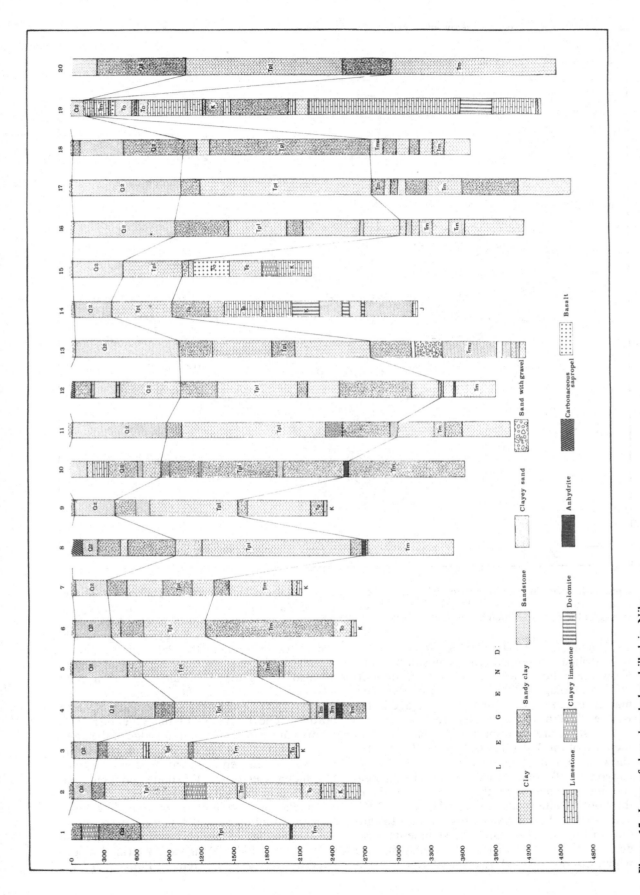

Figure 15. Logs of deep boreholes drilled in Nile delta. Numbers refer to list numbers in Table A-1.

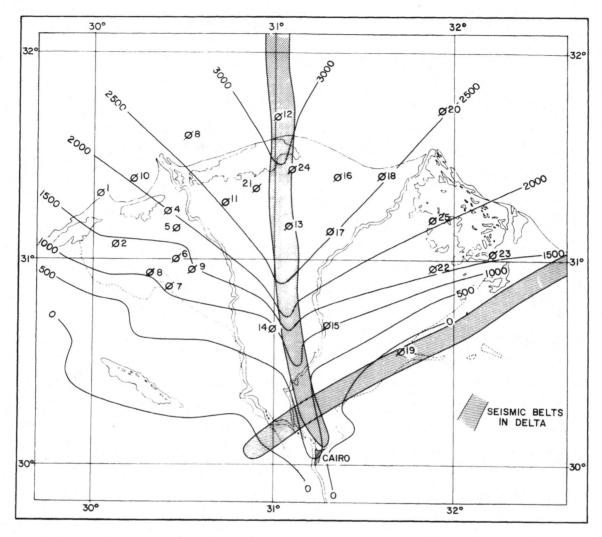

Figure 16. Structure contour map of Paleonile sediments. Numbers refer to wells, the locations and stratigraphic data of which are given in Table A-1.

1973; and Chapter III). The logs of many of the wells drilled in the northern reaches of the delta show the influx of a brackish water foraminiferal assemblage following on top of the open marine assemblage of early Pliocene age in the Kafr el-Sheikh formation (Figure 15).

The structure contour map constructed on the base of the Paleonile sediments (Figure 16) shows a bottom topography in which the Eonile canyon is still obvious. By the end of Paleonile riverine sedimentation the canyon was filled, the South Delta block was overlapped and the delta surface became a slope with a northward dip. In fact, all post-Paleonile sediments were deposited along a northward sloping surface which was not affected by the underlying struc-

ture. Figure 17 gives the isopachous contours of the Paleonile sediments in the delta region. The sediments in the south Delta block are in the range of 250–500 m, whereas in the Northern Delta embayment the Paleonile sediments assume a thickness approaching 1000 m. Figure 17 also shows that the Paleonile had a delta whose ouline extended far beyond that of the modern delta along both its eastern and western borders. The thickness of the Paleonile sediments ranges from 44 m on the edge of the delta in Hosh Isa to 1248 m in Abu Madi situated in the North delta embayment. Table II-4 gives the thicknesses of the Paleonile sediments encountered in the delta boreholes.

The Paleonile sediments are exceedingly uniform in lithological and mineral composition. An examination of four clay samples from these sediments in Sidi Salem well #1 (kindly supplied by the Egyptian General Petroleum Cor-

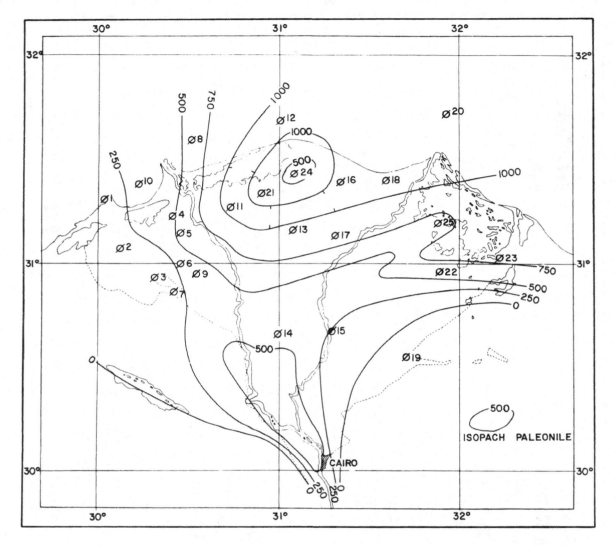

Figure 17. Isopach map of Paleonile sediments. Numbers refer to wells, the locations and stratigraphic data of which are given in Table A-1.

poration) shows that the clay is composed mainly of montmorillonite with little kaolinite and some accessory minerals including quartz, biotite, muscovite, pyrite, epidote, and zircon. Minor feldspars and pyroxenes are recorded from the topmost sample. Save for the presence of minor interbedded sand beds, especially noted in the upper levels of the formation, the Paleonile sediments form an almost solid clay unit. Phillip and Yousri (1964) describe the mineralogy of the sand layers which intercalate this unit in the Abu Sir well near Cairo and state that about one-half of the heavy mineral fraction seperated is made up of opaques. Epidote and various accessory minerals such as zircon, kyanite, andalusite, staurolite, and others make

up the other half of the heavy mineral fraction. No pyroxenes such as those characterizing the modern Nile sediments (Shukri, 1950) are found in these sand layers. This point to a source that is different from that of the present Nile.

Surface Exposures

The Paleonile sediments crop out along the footslopes of the bounding cliffs of the present major wadis that drain into the southern reaches of the Egyptian Nile, indicating that these wadis were not only in existence when the Paleonile started flowing but that they also represented major tributaries of this river. The logs and locations of sections measured in some of these tributaries are given in Figures 18 and 19. Table A-3 lists these sections and gives their elevation and stratigraphic data.

The most important of the tributaries of the

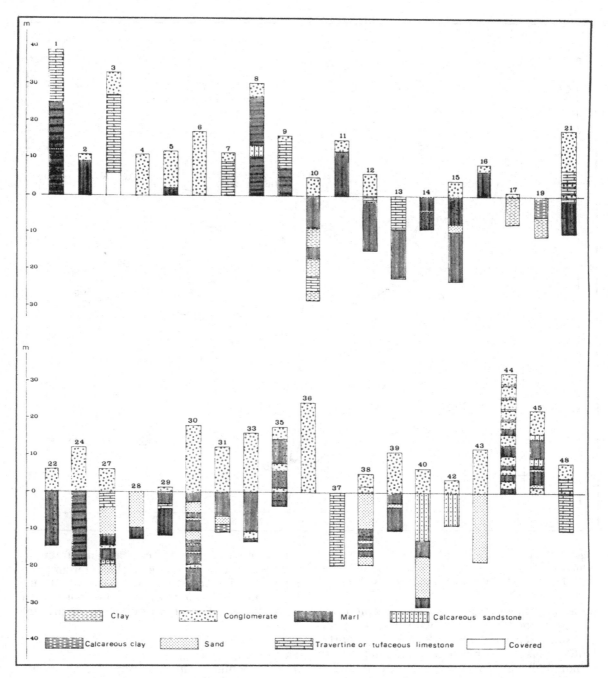

Figure 18. Logs of measured sections (across facing pages) of Paleonile (Tplu) sediments and Armant and Issawia formations, the locations of which are given in Table A-3.

Paleonile system are those which drain the area of the Eastern Desert to the east of the Qena bend. Here, a number of wadis debouch into the Nile via the impressive Wadi Qena which runs in a north–south direction for two degrees of latitude from the Galala plateau to the city of Qena. The cliff which bounds the western side of Wadi Qena is drained mainly by three major tributaries, Aras, Shahadein, and Gurdi all of which deeply dissect the cliff. The cliff to the east is made up of two groups of flat-topped and geologically identical hills, the Abu Had and the Surai-el-Gir complex. The latter lies almost due east of the city of Qena. Several wadis join Wadi Qena from the east, the most important of

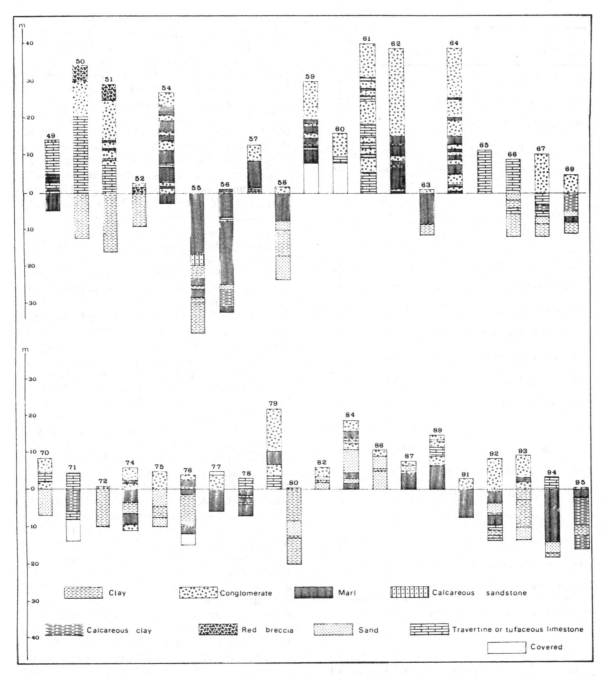

which are the wadis Qreiya and Um Sulimat. To the south of Wadi Qena several northeast–southwest wadis drain the impressive cliffs of the Eastern Desert. Of the important wadis of this stretch are the wadis Matuli, and Madamud which open up opposite Qus and Luxor, respectively.

The type section of the Paleonile is in Wadi Madamud (Figure 20) where a 37.5-m thick section of fine grained silts and clays has been measured. It consists of a lower clay unit of chocolate brown color followed by a bed of alternating laminae of fluvial fine-grained sands and silts and topped by a thin bed of grayish-brown calcareous clay. The section is non-fossiliferous and is topped unconformably by beds of the Armant formation, a poorly sorted torrential deposit (see below).

Along Wadi el-Surai, opposite Qena (Figure 21), the Paleonile sediments are in the form of

Table II-4.

No.[a]	Well	Top Tplu (m)	Bottom Tplu (m)	Thickness (m)
24	Abadiya #1	1169	1723	554
16	Abu Madi #1	966	2214	1248
10	Abu Qir #IX	832	1076	244
12	Baltim #1	978	2112	1134
17	Bilqas #1	1000	1986	986
4	Busseili #1	958	1440	482
6	Damanhour, S. #IX	372	629	257
7	Dilingat, N. #IX	309	495	186
18	el-Wastani #1	1009	2200	1191
3	Hosh Isa #IX	272	316	44
9	Itay el-Barud #1	370	829	459
2	Kafr el-Dawar #IX	195	308	113
13	Kafr el-Sheikh #1	975	1880	905
5	Mahmudiya #IX	667	1229	562
25	Matariya #1	1001	1651?	651?
15	Mit Ghamr #1	483	663	180
23	Qantara #1	735	1450	715
21	Qawasim #1	807	1760	953
20	Ras el-Barr #1	1041	2115	1074
8	Rosetta #1	923	1560	637
22	San el Hagar #1	810	1100	290
14	Shebin el-Kom #1	358	635	277
11	Sidi Salem #1	877	2000	1123

[a] Number of well as it appears in Table A-1 and in all figures.

two clay layers separated by a thin layer of fine-grained sediments topped uncomfortably by the torrential deposits of the Armant formation.

Perhaps the largest of the tributaries of the Paleonile is the Kharit-Garara branch on the sides of which Pliocene fluviatile sediments seem to exist. These Paleonile sediments fringing the present-day course of the Kharit-Garara wadis of the south Eastern Desert of Egypt were identified by photogeological methods (see map in Hunting Geology and Geophysics, 1967). Further work is needed before a full

description of this important tributary of the Paelonile can be given and the following remarks must be regarded as preliminary.

Wadi Kharit (Figure 23), one of the present great trunk wadis of Egypt, has its principal head at Gebel Ras el-Kharit on the main Nile-Red Sea watershed at lat. 24°09′N and long. 35°E. Pursing a course the prevailing direction of which is a little north of west, and collecting the drainage from numerous great wadis on its way, it debouches in Kom Ombo plain and reaches the Nile at the same point as another great wadi, Shait, at lat. 24°35′ N. The length of the main channel is over 260 km, and that of its tributaries probably more than 20 times as long. It drains an area of more than 25,000 km². Its average fall is about 2 km, but in its lower reaches its gradient is less than half this amount. The principal tributaries of Wadi Kharit are wadis Natash, Antar, Khashab, Abu Hamamid, and Garara. This last wadi together with wadis Timsah and Ghadarib are great feeders and are in direct continuation with the main trunk of Wadi Kharit. Together they run in a great tectonic graben that extends from the Kom Ombo plain down to Gebel Hodein.

The Paleonile sediments of the Wadi Kharit-Wadi Garara complex occur as enormous terrace embankments about 10–12 m high on either side of the present water channel. They are made up of well-stratified, alternating thinly bedded friable sand and sandy clay, ending abruptly against the granite and schist of the hill masses. They contain ferruginous crusts and show sun cracks in places. The material is fine and well stratified and must have been deposited under different climatic conditions from those prevailing at present. The deposits are indeed very similar to the Pliocene thinly bedded sediments that are described by Chumakov (1967) as occurring at depth in the boreholes of the Aswan region. The ferruginous crusts are exclusively in the ferrous form indicating reducing conditions. This can be explained by deposition of a river that had its source in a moist region with effective vegetation cover and little or no torrential runoff, so that stream discharge included little or no coarse detritus.

The Kharit-Garara Paleonile sediments run in a northwest direction and end in the Kom Ombo graben. Another similar deposit trending in the same direction fringes the tectonic graben of Wadi Um Sellim which opens up at present to the south of Idfu. The deposits here are more sandy than the Kharit-Garara sediments. Whether this wadi during the Pliocene was in continuation with the Kharit-Garara drainage or not is yet undetermined, but it is certain that

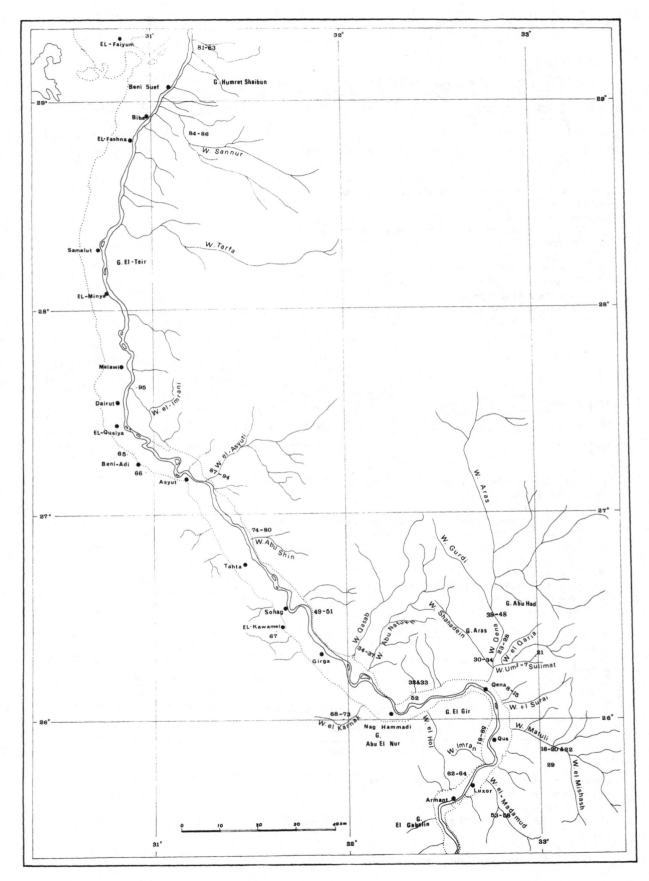

Figure 19. Key map showing locations of measured
sections.

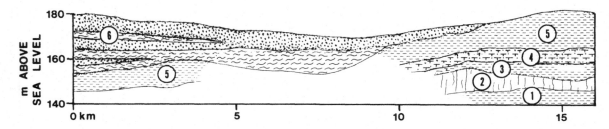

Figure 20. Section along Wadi el-Madamud, opposite Luxor, Upper Egypt.
  Armant formation
  6. Alternating beds of coarse sand, conglomerate and marl
  Paleonile sediments
  5. Brown shale with partings
  4. Fine-grained, nonfossiliferous sandstone
  3. Gray shale with gypsum
  2. Fine-grained sandstone and silt, light brown
  1. Shale, choclate brown

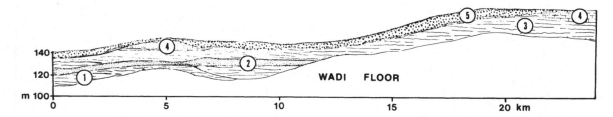

Figure 21. Section along Wadi el-Surai, opposite Qena, Upper Egypt.
  Armant formation
  5. Hard conglomerate made up of limestone cobbles, lying unconformably over Paleonile sediments
  Paleonile sediments
  4. Tufaceous limestone
  3. Upper clay, calcareous, dark brown
  2. Fine-grained sand
  1. Lower clay

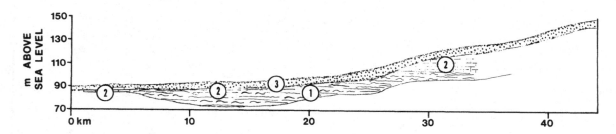

Figure 22. Section along Wadi el-Assiuti, opposite Assiut, Upper Egypt.
  Armant formation
  3. Coarse-grained sand and gravel
  Paleonile sediments
  2. Tufaceous limestone
  1. Clay, calcareous, choclate brown in color

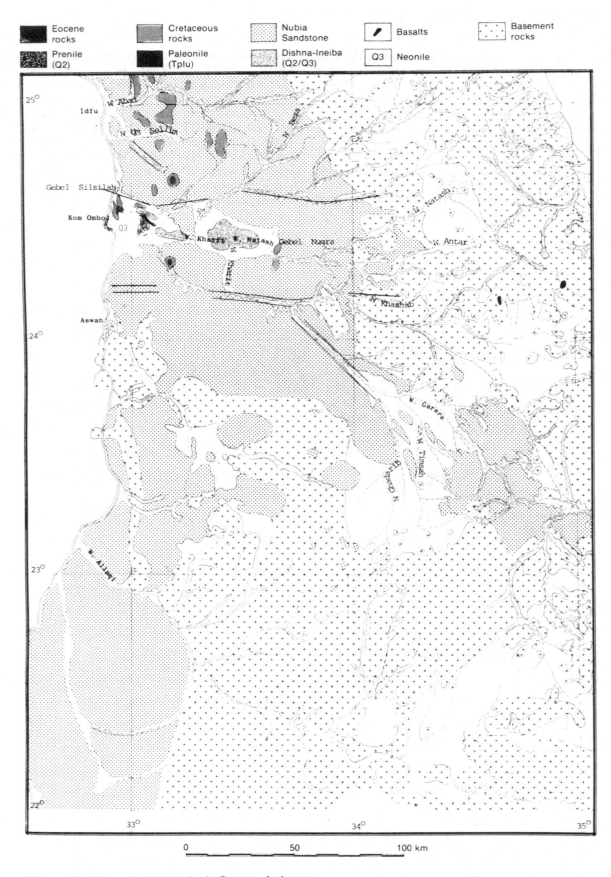

Figure 23. Map of Wadi Kharit-Garara drainage basin.

the northwest—southeast grabens of wadis Um Sellim and Kharit-Garara are older than the east—west Wadi Shait-Silsila, Wadi Natash, and Abu Sibeira-Allawi-Khashab faults which control the Kom Ombo graben. Figure 23 sketches the Kharit-Garara drainage and the major tectonic lines of the region.

The Paleonile Deltaic Deposits

The fluvio marine and deltaic beds of the Wadi el-Natrun area (Gar el-Muluk formation) seem to represent the western peripheral deposits of the Paleonile delta in its early phases. The Wadi Natrun fluviomarine beds have attracted the attention of authors since the discovery within them of fossil mammals by Stromer von Reichenbach in 1889. The area has since then been visited by numerous scholars and good descriptions of the classical Gar el-Muluk section are given in Lyons (1906), Blanckenhorn (1901), Andrews (1902), and Stromer von Reichenbach (1902a).

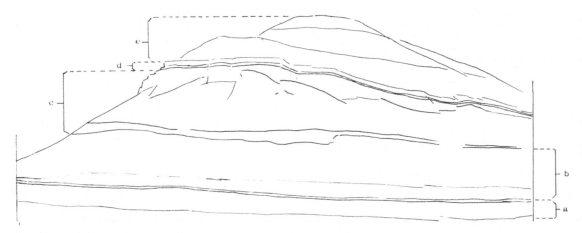

Figure 24. Section at Gar el-Muluk, Wadi Natrun.

The last author describes the vertebrate faunas as well as the stratigraphic relationships of the fossil vertebrate-bearing beds within the section. A review of the earlier work on the vertebrates found in this region and a description of additional materials are to be found in James and Slaughter (1974).

The exposure at Gar el-Muluk (Figure 24) forms a prominent topographic feature in the flat surrounding terrain of the Natrun depression. The hill is capped by about 14 m of alternating gypseous clay, sandstone, silt, and limestone beds of both marine and brackish water origin (beds "e" by Stromer von Reichenbach's lettering). These beds contain, in addition to crocodilian and fish remains, a late early Pliocene ostracod fauna. Below this unit is a 1-m thick resistant sandstone bed (bed "d") followed underneath by slope-forming carbonaceous sandstone, clay, and limestone beds 8 m thick (beds "c"). Underlying this are the main vertebrate-bearing beds which include 6 m of alternating sand and clay layers with cross-bedded channel-fill deposits of sand and gravel (beds "b"). These last beds seem to be of freshwater origin. *Hipparion, Hippopotamus,* and *Hippotragus* spp. are among the vertebrates described from this unit. The fauna seems to indicate a late lower Pliocene age (James and Slaughter, 1974).

e	1.60 m	Glauconitic gypseous sand; occasional chert pebbles	
	2.00	Dark colored shale	
	.10	Limestone band rich in ostracod shells and fish bones	
	10.00	Green shale from which one skull of crocodile is reported	
d	1.00	Calcareous sandstone with ostracod shells, forming ledge	
c	8.00	Green and dirty gray carbonacores shaly sandstone	
b	6.00	Alternating sand and clay layers, cross-bedded with vertebrate remains: *Hipparion, Hippopotamus* spp. etc.	
	2.60	Gray gypseous shales with bones	
a	.50	Gray shale with *Ostrea cucullata*	
	1.50	Black carboneceous shale with plant remains	

The Pliocene beds here show that the early Pliocene marine incursions with *Ostrea cucullata* antedate and interfinger the fluviatile deposits of the Paleonile.

The Wadi Natrun fluviomarine beds of the Pliocene exposure thus seem to represent the oldest beds deposited by the Paleonile which began with the retreating seas of the early Pliocene. The river then seems to have overstepped this area to build a large delta toward the north. Figure 17 shows that the Paleonile sediments increase in thickness progressively northward. It seems reasonable, therefore, to believe that these sediments were desposited by a prograding river which built up a progressively advancing delta, the northern limits of which within the Mediterranean cannot be at present delineated. The Nile cone fans out underneath the Mediterranean waters. It covers the continental shelf and slope and extends across the continental rise to the long and narrow Herodotus Abyssal plain (Figure 25). The area of the Nile cone is about 125,000 km², almost four times the size of the modern continental delta. The western part of this cone (the so-called Rosetta fan) is the more extensive stretching across the continental slope into the cone. The slope itself is indented by the Alexandria submarine canyon off the modern Rosetta branch. The canyon begins a branching network of channels dissecting the cone. The eastern part (the so-called Damietta fan) is a thin wedge lying to the west of the Levant platform. The smallness of this fan leads Ross and Uchupi (1977) to disregard the Damietta fan and map the entire cone as one unit rather than two (compare Emery, *et al.,* 1966).

The sediments of the Nile were carried far beyond the Nile cone. The work of the Deep Sea Drilling Project Cruise Leg 13 (Ryan *et al.,* 1973a) shows that the Mediterranean ridge is made up of a sequence of terrigenous muds, sands and sandstones more than 500 m in thickness, sparsely intercalated with pelagic marl oozes of Quaternary age. Sedimentary textures and primary bedding structures suggest that the terrigenous layers of this formation were deposited for the most part by turbidity currents; and mineralogical investigation of both the fine- and coarse-grained detrital components suggest a Nile River provenance. The southern flank of the Mediterranean ridge is an uplifted and deformed wedge of basinal sediments, previously deposited on a once extensive abyssal plain seaward of the Nile cone. Assemblages of foraminifera and dated sequences of sapropelitic muds and volcanic tephra in the superficial layer of pelagic sediment (which unconformably overlies the terrigenous sediments) indicate that uplift of the sea bed isolated this part of the ridge from terrigenous

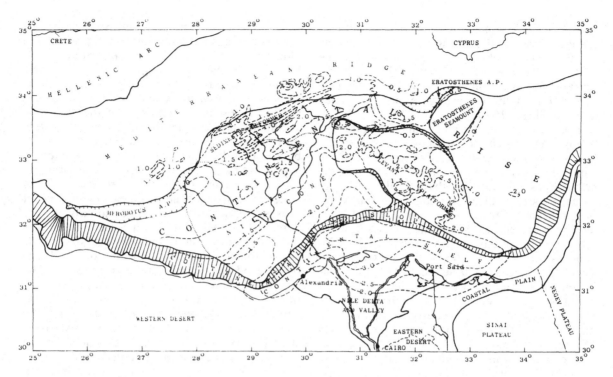

Figure 25. Physiographic provinces of coastal region and adjacent sea floor. Dashed lines are isopach contours of Pliocene-Quaternary sediments above acoustic basement measured in seconds. Modified from Ross and Uchupi (1977).

deposition of Nile origin sometime around half million years ago. It is, therefore, feasible to believe that the Paleonile contributed a large part of the sediment of the Nile cone and the Mediterranean ridge and that the deformation of the ridge took place during the highly seismic early Pleistocene interval (see below).

The thickness of the deltaic sediments since the beginning of river sediment contribution to the Mediterranean varies from one place to another. The thickest section is over $3^{1}/_{2}$ km in thickness and lies along the Continental Shelf off the delta (Figure 25). The isopachous contours of the Pliocene-Quaternary sediments off the Nile delta indicate that thick sections run along a median line which coincides with the offshore extension of the seismic axial zone of the mid-delta (Figure 16). Considering the average thickness of the sediments, it is possible to estimate the volume of the entire deltaic sediment to be 350,000 km³. This figure is larger than the 140,000 km³ given by Harrison (1955) based upon an interpretation of gravity anomalies in the eastern Mediterranean. Ross and Uchupi (1977) give a figure of 387,000 km³ for the volume of sediment of the Nile delta. Of

the estimated 350,000 km³ of the Nile delta cone sediments, it can be safely stated that about 20% of this detritus resulted from accumulation by the Paleonile system. More than 20% of the thickness of the late Neogene and Quaternary sediments encountered in the boreholes drilled in the delta belong to this unit.

Possible Sources of the Paleonile
The uniformity of the Paleonile sediments, their mineralogy, and their fine-grained lithological composition suggest that the Paleonile had its sources in moist areas with effective vegetation cover. Indeed, southern Egypt must have enjoyed a wet climate during the Pliocene. It is interesting that the flow of the Paleonile coincided with the great first global cooling (see discussion in Chapter 3), which must have brought climatic changes to tropical and equatorial Africa. The distribution of the Paleonile sediments suggests that the wadis draining the southeastern massif of Egypt formed the main source of this river (Figure 26). The fauna of the Paleonile sediments show no tropical or African elements. The many Central African freshwater molluscs frequently found in the late Eocene, Oligocene, and early Pliocene fluviomarine deposits of Egypt are completely lacking in the late Pliocene of the Nile Valley. The genera *Melania, Melanopsis, Vivipara,* and *Neritina* found in the Paleonile sediments are rather Mediterranean in aspect (Blanckenhorn,

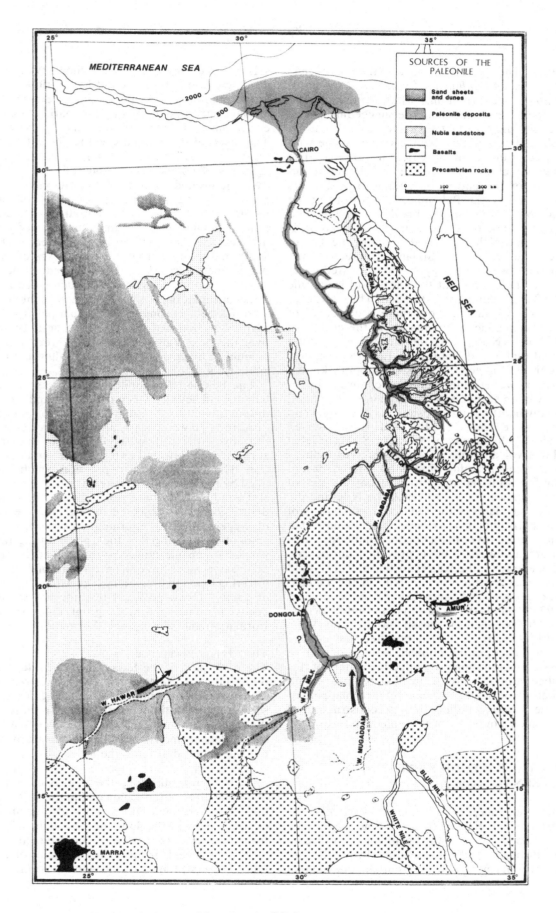

Figure 26. Map showing possible sources of Paleo-

1921). The typical tropical African species, such as *Ampullaria, Lanistes, Cleopatra, Spatha,* and *Aetheria,* which consitute the fauna of the sediments of the succeeding rivers, are missing. This fact seems to suggest that the Paleonile was not connected to a drainage with sources from Central Africa or Ethiopia. The Egyptian climate must have been humid to account for the enormous volume of sediments deposited by the Paleonile. The history of the equatorial Nile and its breaking through into Egypt is unknown. No detailed information is available which would help in dating the flow of the drainage of the Sudan basin across the Nubian swell into Egypt and the Mediterranean. There are great thicknesses of lacustrine and fluvial sediments in the Sudan basin from at least the Cretaceous, but there is no record of pre-Pleistocene sediments along the course of the Nile in Nubia. It is, therefore, likely that the equatorial Nile did not reach Egypt except during the Pleistocene. At this stage of our knowledge, however, one cannot preclude the possibility of an earlier breakthrough during a period of downcutting and erosion via the large and impressive wadis el-Milk and Hawar which drain the highlands of Western Sudan (Gebel Marra volcanic massif which receives today a mean annual rainfall of 80 mm). These seasonally fed wadis emanate today from this massif and their waters fail to reach the Nile and die out in the plains of the sand-covered country of Northern Sudan. It is feasible to assume that under more favorable climatic conditions, such as are assumed to have prevailed during late Pliocene time, the waters of these wadis could have reached the Nubian Nile in the Debba-Dongola reach. Andrew (1948) and Whiteman (1971) claim that there is an uplift of the country between the massif and the Nile (Wadi el-Milk-Sodiri axis). It is possible that in later times this uplifting coupled with a period of lesser precipitation prevented the waters of this subequatorial source from flowing into Egypt. There is no doubt that the detailed mapping of these regions will help in the reconstruction of the paleogeography of the Paleonile basin.

II-4. THE PALEONILE/PROTONILE INTERVAL (Tplu/Q₁)

The Paleonile/Protonile interval covers the period which elapsed from the time the Paleonile stopped to flow to the beginning of the breaking through of the succeeding Protonile into Egypt. This interval was essentially one of great

seismicity during which the climate was exceedingly arid. This period of hyperaridity was interrupted by a short pluvial, here termed the Armant pluvial, during which semiarid climatic conditions with winter season runoff prevailed. The result of this pluvial was the deposition of locally derived detritus in the form of alternating beds of conglomerate and sands or marls over the eroded surface of the Paleonile sediments. The beds today choke many of the mouths of the side wadis that drain into the Nile. Thicknesses of as much as 40 m of this material are noted in the Armant area to the south of Luxor. During the end phases of this pluvial (here termed the Issawia) bedded horizontal or slope travertines were deposited and tufaceous materials cementing the Armant beds were formed. During the latter part of the Issawia time interval great seismic activity set over the Nile Valley. Thick talus breccias accumulated along the slopes of the valley and the side wadis.

The Paleonile/Protonile interval was one in which wind activity was most active; and it was during this interval that great physical changes occurred to the landscape of Egypt. The vegetative cover was destroyed and many of the land surfaces were lowered as a result of wind deflation. The Wadi Natrun depression which was filled by the Paleonile sediments, relics of which still stand in the form of isolated hills within the depression, was subjected during this interval of hyperaridity to enormous lowering. The amount of lowering that took place during this particular episode cannot be determined, for the depression must have been the site of further lowering by wind erosion in succeeding periods of aridity. The deposits of this interval will be dealt with under the headings of The Armant Formation and The Issawia Formation.

The Armant Formation

The early part of the hyperarid Paleonile/Protonile interval seems to have been characterized by winter torrential runoff. During this episode conglomeratic deposits similar to those of the recent past (marked Q_w on the accompanying geological maps) accumulated in many of the wadis that drain into the Nile.

The conglomeratic deposits of the Armant formation crop out along the footslopes of the bounding cliffs of the valley and in the deltas of the wadis that open into it. They also occur at depth in many of the floors of these deltas. The formation has not been recognized in the boreholes drilled in the valley or delta of the Nile. The distribution of the Armant formation

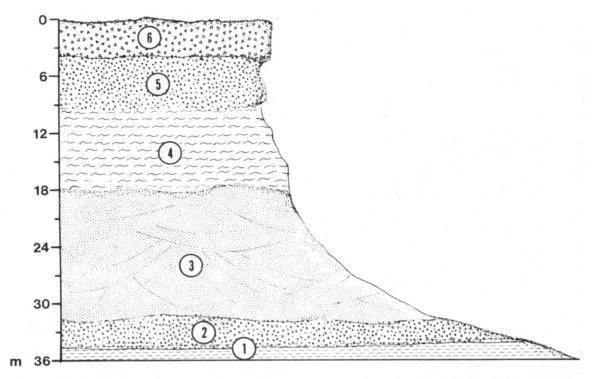

Figure 27. Section of the cliff opposite Abydos, Upper Egypt.

6. Cemented red breccia made up of angular chert and quartz pebbles (Brocatelli)
5. Conglomerate cemented with tufaceous material
4. Marl, brown with gypsum veins
3. Sand with clay interbeds, locally cross-bedded and with clay concretions
2. Sand, coarse-grained with many pebble bands
1. Bedrock (Esna shale)

can be seen in the accompanying geological maps where it skirts the cliffs that hem the valley or the wadis that drain into it and makes the floor of many of the open deltas of these wadis. Some of the finest exposures are found at the footslopes of the towering cliffs of the plateau in the Thebaid hills (Figure 12) on both the east and west banks, and on the east bank of the Minia-Beni Suef stretch where later erosion exposed sections of this formation composed of layers of conglomerate made up of the coarse disintegration products of the surrounding plateau, washed and rounded by tributary streams. Away from the cliffs, the fine-grained calcareous component of this formation becomes increasingly conspicuous. These fine-grained sediments and marls composed largely of pulverized material of Eocene limestone are cemented in some places (such as in the Valley of

the Kings, Luxor) by later action of springs into fine-grained hard rock.

To the north of Abydos, Sandford (1934, p. 24) describes a section of the Armant formation which is topped by the red breccia of the Issawia formation. The section (Figure 27) is made up of an upper 10-m- thick laminated marl with minor sand and fissile shale interbeds overlying a 15-m- thick cross-bedded sand layer with root drip and balls of brown mud. This in turn overlies a bed of fine conglomerate and quartz sand. The thickness of the lower bed is 5 m. It rests unconformably over Esna shale bedrock. The minerology of the sand beds of this section is worked out by Hassan (1976). The sands are devoid of augite, rich in epidote, and have a totally different composition from the Qena sands (see below). In Beni Adi, Manfalut district, the Armant and Issawia beds overlie inconformably the Paleonile chocolate marls (Figure 28). The Armant is made up of alternating pebble beds and marls followed on top by horizontally bedded travertines with tube marks (reeds).

The most complete section of this formation is in Wadi Bairiya, Luxor district, which opens into the Nile near Armant (Figure 29). Here, the formation consists of beds of conglomerate made up of locally derived Eocene pebbles of limestone and chert of different sizes, alternating with thick marl beds which have numerous

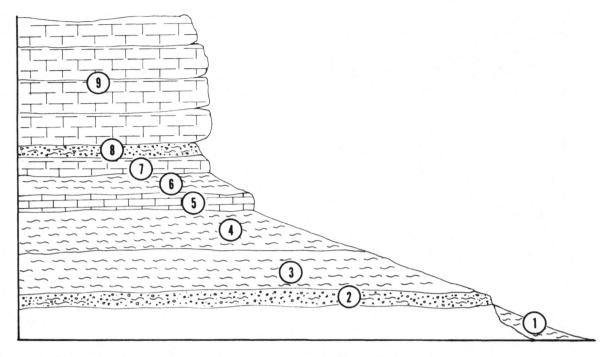

Figure 28. Section of cliff west of Beni Adi, south-west Manfalut, Upper Egypt.
9. Horizontally bedded travertine
8. Flint and quartz pebbles in calcareous marl
7. Travertine
6. Marl, green with white specks
5. Limestone, porous
4. Marl, hard, light gray
3. Marl, hard, light gray
2. Pebble bed in marl made up of rolled flint, quartz, and siliceous pebbles
1. Marl, chocolate brown

pebble intercalations. Sandford and Arkell (1933, Plate III, A and B) give a photograph of this section and a good description of it under their "Pliocene Gulf Deposits." The Bairiya section is also the subject of the study of Coque and Said (1972) and Biberson, *et al.* (1977) where "evolved Pre-Acheulian" implements are claimed to have been found in some of the gravel beds of the Armant formation of this locality.

To this formation belong also the east–west gravel-filled inverted channels that seem to have drained the elevated Faiyum region and opened in the Rus channel of the Nile. Sandford and Arkell (1929) are the first to delineate these channels and a good description of them is given in their work and in Pfannenstiel (1953). These authors interpret these channels as inter-digitating the marine Pliocene sediments in the Nile Valley. The reexamination of these deposits, however, leads the present author to believe

that the sediments follow unconformably on top of the marine Pliocene sediments.

Deposits that belong to the Armant formation are recognized by Sandford and Arkell (1933, 1939) where they are classified among the "Pliocene Gulf Deposits", and by Blanckenhorn (1901, 1921) where they are separated as "Diluvium II deposits."

The composition of the rock fragments that make up the deposits of the Armant formation differs from place to place, being on the whole of calcareous nature owing to the derivation of most of them from the Eocene limestone plateaus that tower the valley along most of its middle course; but in some areas the fine-grained rock is made up of the products derived from the disintegration of the Cretaceous to Paleocene Dakhla and Esna shales. These can be easily recognized because they are extensively quarried for use as fertilizer by the local villagers. In other areas, such as Wadi Um Sulimat, a tributary of Wadi Qena (Figure 30), the Nubia sandstone and the phosphate formation give the rock a phosphatic or arenaceous composition. In many areas, such as in Beni Suef, the gravel ridge made up of loose siliceous to dolomitic gravels is quarried for local construction. In Beni Hassan, Minia province, a 15-m thick section of thinly bedded coquinal limestone of reworked *Nummulites* shells from the nearby Eocene is recorded. In the Qena and Ballas regions the calcareous silts famous since ancient Egyptian times were and still are used

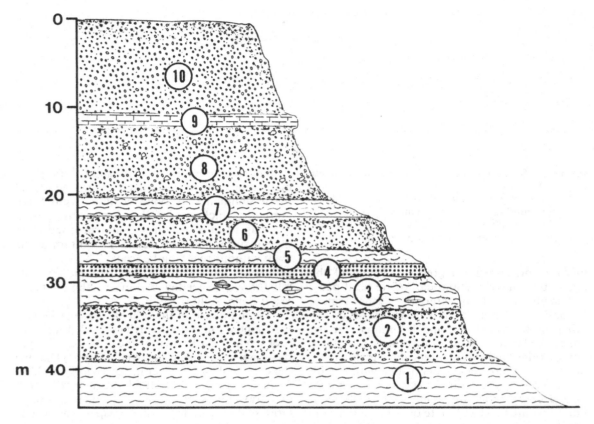

Figure 29. Section at the side of Wadi Bairiya, opposite Armant, Upper Egypt.

10. Conglomerate made up of gravels of limestone and chert, few pockets of sandy material
9. Travertine, horizontally bedded, brown
8. Conglomerate made up of large (5–50 cm diameter) boulders cemented by tufaceous material
7. Marl, gray to brown
6. Conglomerate made up of 1–10 cm diameter pebbles of chert and siliceous limestone embedded in a matrix of tufaceous material
5. Marl, gray to brown
4. Coarse-grained quartz sand
3. Marl with calcareous concretions
2. Conglomerate made up of 1–10 cm diameter pebbles
1. Marl, gray

as raw material for the famed Qena potteries of Upper Egypt.

Recent studies of these "Pleistocene" clays using techniques of neutron activation (Tobia and Sayre, 1974) show that the composition of the clays differs from one locality to another depending on the shale bedrock from which these clays were derived. The Qena material for example is more or less similar in composition to the outcropping nearby Paleocene Esna shale. They have less iron, cesium, europeum, and thorium and more barium and chromium than the clays from Maadi, near Cairo, which were derived from the outcropping nearby upper Eocene Maadi formation.

The Armant formation lies unconformably over the Paleonile or older sediments. The thickness ranges from a few meters to more than 40 m. Its absolute elevation ranges from 120 to 230 m above sea level. On the whole, the elevation increases as one moves from the mouth to the source of the side wadi, indicating a steep gradient (1:200 in case of Wadi Qena) and formation after an intense period of erosion. Table A-3 gives the thickness and absolute elevations of the sections of the Armant formation measured in the field, while Figure 18 depicts the logs of these sections.

One of the characteristics of the Armant formation which distinguishes it from later modern wadi conglomerates is the yellow red color which it assumes as a result of the presence of iron oxide cement within its particles. Most of the Armant occurrences form rolling surfaces. The top gravel layers (up to 2 m deep from the surface) have, in between the pebbles, sandy loam which is reddish brown in color (3 YR 4/4 moist–4/7 dry) and is rich in lime with nu-

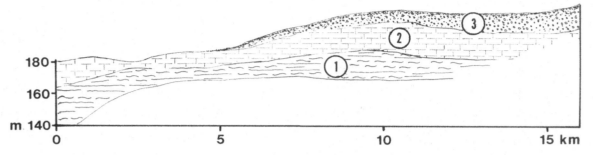

Figure 30. Section along Wadi Um Sulimat, a tributary of Wadi Qena.
3. Conglomerate, made up of limestone pebbles cemented in a tufaceous matrix
2. Tufaceous limestone, porous
1. Marl with intercalations of fine-grained silt

merous soft gypsum aggregates. Many of the gravel beds have calcium carbonate hard pans.

The lithological characteristics of the Armant formation are not different from those of another more recent but much thinner deposit which is also attributed to torrential activity of winter rains and cloudbursts during the Neolithic. This more recent deposit, together with other scree, appears on the accompanying geological maps under the symbol Q_w (subrecent fanglomerates). The climate of the Armant time seems to have been similar to that of the Neolithic, but probably of much longer duration and perhaps of more frequent rains.

It must be assumed that the arid conditions that set over Egypt at the advent of the Pleistocene caused the Paleonile to slacken or stop flowing into Egypt. During the late Pliocene (Paleonile time) the climate seems to have been wet throughout the year favoring the development of a thick vegetative cover which protected the land from intensive erosion. Only fine-grained sediments were washed away and transported by the Paleonile. As arid conditions set over Egypt during the early Pleistocene, the vegetative cover was destroyed and intensive mechanical weathering took place. During the Armant pluvial which interrupted this arid phase, sporadic but frequent winter cloud bursts moved the weathering materials not far from their source.

The depth of erosion of the Nile Valley sediments in Armant time cannot at present be estimated, but it was probably in the range of 40 m judging from the difference in elevation of the Paleonile sediments in the thalweg and in the sides of the Nile Valley. The Armant (ephemeral) master stream seems to have first cut into the Paleonile sediments and then built up its bed in a course that is yet to be defined. The

volume of the sediment deposited during Armant time can be visualized from a study of these sediments in Wadi Sannur, Beni Suef. Here the thickness of the Armant formation exceeds 15 m of which 9 m are recorded from the subsurface in a pit dug in the bottom of the wadi without reaching bedrock. The volume of sediment deposited by the Armant torrents must have exceeded 3 km³ in the delta of Wadi Sannur alone.

The end phase of this pluvial was marked by the deposition of travertines and tufaceous rocks. The best-known of these travertines are those which occur in Issawia (Figure 31) along the east bank of the river in the vicinity of Akhmim. The famous 7-m thick, horizontally bedded travertines of Issawia are quarried on a large scale for their peculiar property as water-resistant rocks. The Issawia travertines are separated from the underlying chocolate-brown clays of the Paleonile by a ¹/₂-m thick bed of conglomerate made up of cemented rounded to subrounded limestone pebbles. The travertine itself, although porous, is very hard, easily dressed and has no tendency to split or break, but can be worked in any direction and cut into any shape. The specific gravity is 2.4 to 2.5, while crushing strengths reach up to 188 kg/cm². It is almost exclusively made up of pure $CaCo_3$ (99.60%) which also deposited in thin imbricating laminae around algae and plants which thrived in pools formed during the Armant pluvial. Although the rock itself is creamy-white in color, the cavities in the limestone are often lined with red or yellow ocher. Mineralogically, the rock is made up of fine-grained microcrystalline calcite which is cloudy gray to white in color. The mineral is in the form of very fine anhedral to euhedral crystals, and there are few and scattered quartz grains.

Travertines that belong to this episode are known in many places along the slopes of the Eocene limestone cliffs of middle Egypt. The largest occurrence is in Naga Hammadi west where the travertines form an unbedded mass of vesicular rock that follows on the slope of the cliff. They seem to owe their origin, as in the

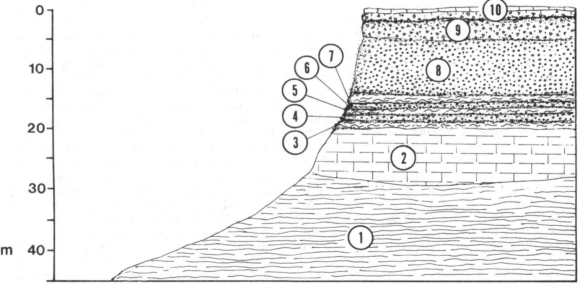

Figure 31. Section at Issawia quarry, opposite Akhmim, Upper Egypt.

10. Weathered bed of red breccia covered with slabs of wind-faceted siliceous limestone
9. Red breccia made up of angular, siliceous limestone pebbles embedded in a matrix of red-brown muds (Brocatelli)
8. Conglomerate made up of siliceous limestone pebbles embedded in a tufaceous matrix
7. Marl, brown
6. Conglomerate as in 8
5. Marl, brown
4. Conglomerate as in 8
3. Marl, brown

2. Travertine, horizontally bedded, hard and with plant remains
1. Clay, grayish-brown with gypsum specks in cracks.

case of Kharga oasis slope travertines, to the accumulation of the carbontes of the oozing waters spilling from the cliff around reeds whose roots must have taken hold on the cliff.

Other travertines that belong to the same episode are found west of Beni Adi and Deir el-Muharraq. In the Valley of the Kings, around Luxor, the conglomerate and marl beds of the Armant formation are cemented by tufaceous materials that seem to have formed as a result of the evaporation of the $CaCo_3$-bearing waters that percolated through them.

Since travertines and other tufaceous materials form at the end phase of a pluvial period, travertines could be of different ages. Our work, however, shows that the end phase of the Armant pluvial is the episode *par excellence* of travertine formation in the Nile Valley, as most of these travertines are associated with sediments of the Armant formation.

The Issawia Formation

The Issawia formation follows on top of the Armant in many places in the Nile Valley. It is made up of massive rubble breccias which are topped in many places by the characteristic hard red breccias which are quarried for ornamental use in many places along the Nile. The type locality of this formation is in the Issawia rhomboid reentrant along the east bank of the river in the vicinity of Akhmim. The section (Figure 18, section 51) is made up of a 15-m thick massive bed of breccia made up of angular limestone pebbles overlying unconformably the thick travertines of this locality. It is characterized by poorly sorted angular pebbles, boulders, and occasional large blocks of stones. Usually, however, most of the pebbles are in the size range of 4–6 cm or less which gives the land surface a rather smooth appearance. The beds occupy the piedmonts of the nearby cliff, and their slope is rather steep (between 10 and 20%). The breccia was obviously not deposited by running water, but largely by gravity.

The smooth surface of the formation consists of a hard and cemented red breccia, $6^1/_2$-m deep in Issawia, which has been quarried since ancient Egyptian time for use as an ornamental stone. The commercial name of this stone in Egypt today is "Brocatelli." The color of the breccia is due to iron oxides of red-brown color which seem to have filled the pore spaces of the breccia during a period of soil formation in a later episode of humidity; the breccias were later cemented by the evaporating $CaCo_3$-bearing waters that percolated through the pebbles. Red breccias formed during the end phases of

pluvials are noted in Luxor, Abydos, Issawia, Tahta, Wadi Qena, and Wadi Assiut. In Abydos the red breccia is made up of Eocene chert and limestone. It is almost 3 m in thickness and overlies conformably a 7-m thick white conglomerate of finer grain than the red breccia cemented with travertine. When traced laterally the conglomerate bed passes into travertine with inclusions of balls of brown mud (Figure 27).

As previously stated, the massive breccias of the Issawia formation are accumulations due to gravity and mass movements which seem to have been triggered by earth movements which must have affected the valley during the Issawia episode. The fact that most of the pebbles are angular chips indicates that these movements occured during and after a long period of great aridity. Red breccias are recorded by Sandford (1934, p. 34) outside the valley in five areas in minor depressions in the plateau to the west of Minia. These occurrences fall within the range of the map that accompanies this book and are marked as "fanglomerates of older age." All these occurrences lie along northwest –southeast faults that cross the Nile in this region producing the rugged topography of the east and causing several isolated outliers of Eocene to appear within the valley on the west bank of this reach. Basalt dikes occur on both sides of the Nile in the Samalut-Beni Mazar highly faulted region. The age of these basalts is determined by Meneisy and Kreuzer (1974) as Oligocene (28 million years). It is most likely that these faults are of the same age and have been episodically active since that time. The presence of red breccias in these depressions indicates that these transverse northwest faults must have been activated during the Issawia time to produce these massive breccias.

It is interesting that the highly seismic Issawia time interval coincides with the Plio-Pleistocene interval which Ross and Schlee (1973) demonstrate to be the interval of sea-floor spreading resulting in the formation of the axial zone of the Red Sea.

It may be of interest to point out also that formations similar to the Armant and Issawia are recorded from many parts of North Africa (Coque, 1962), and they are assigned to the Villafranchian Aidian stage (Chavaillon, 1964).

II-5. THE PROTONILE (Q_1)

Fluviatile deposits that belong to the third river system that occupied the present Nile basin are in the form of complex gravel, coarse sand, and loamy materials. The best and most extensive exposure of the remains of this river are to be found in the Kom Ombo west bank stretching across Darb el-Gallaba plain from Wadi Kubaniya to the north of Aswan as far as el-Sibaiya. This formation, named Idfu, has its type locality in Wadi el-Qura where extensive work has been recently completed to evaluate the gravels of this region as a possible source of silica for use in the ferrosilicon and phosphorous complex industries.

The Darb el-Gallaba area is almost completely covered by these Idfu gravels, interrupted only where the Nubia sandstone bedrock crops out in some areas. Between these isolated outcrops of bedrock, the Idfu River gravels are present everywhere, more or less eroded by gullies formed in later erosion stages which mostly drain to the present river. Certain outcrops of Nubia sandstone have relics of the Idfu gravels on their tops, which indicates that these gravels must have been at least 30 m higher than their present level estimated to be about 50 to 60 m above the level of the Nile.

The Darb el-Gallaba plain represents the flood plain deposits of a braided competent river which carried mainly gravel over an everbranching system of stream beds. Wadi el-Qura is a small wadi which cuts through the Idfu gravels and opens up near the village of el-Hassayia west, 11 km south of Idfu. Pit #1, to the south of this wadi, is taken as the type section of this formation. The pit was sunk for a total depth of 7 m in one of the ridges which form the relatively higher parts of the rolling plain. At depth $4^1/_2$ m a loose yellow coarse sand layer with occasional quartz pebbles occurs and continues for the remaining $2^1/_2$ m of the pit without reaching the base of the layer. The upper $4^1/_2$ m are made up of a top thin 10 cm layer of cemented conglomerate made up of large (+ 50 mm) well-sorted quartz pebbles, mainly milky white in color cemented in a matrix of medium- to fine-grained red-brown compact sandstone. This is underlain by a gravel bed made up of loose, well-sorted rounded to subrounded quartz pebbles, mainly milky white in color but becoming pale brownish at depth. The whole section is topped unconformably by a surface gravel layer about 10–30 cm thick which is made up of locally derived Eocene and Cretaceous chert pebbles derived from the plateau in the hinterland.

In the hundred wells that were drilled in this area the section is more or less the same. The loose gravel layer ranges in thickness from 2 to

$4^1/_2$ m, and it always rests over a bed of coarse sand which exceeds in some pits 6 m in thickness. The mechanical analysis of the loose gravel layer in pit #1 shows that 60% of the pebbles are in the size range 50–100 mm, 6.7% are in the size range 10–50 mm, 1.5% are in the size range over 100 mm, while 31% are finer than 10 mm in size. Analysis of a hundred other representative samples of this bed shows that more than one half of the gravels are in the size range 50–100 mm (reaching up to 83% in one composite sample); and the rest is distributed in the other ranges, mostly in the finer than 10 mm range. The mineral analysis of the gravels shows that they are overwhelmingly siliceous (92–96%), other igneous rock fragments comprising the remainder of the ingredients. The chemical analysis shows that the percentage of silica may reach 96%.

The topmost conglomeratic layer is cemented by a yellowish red coarse sandy loam (5 YR 4/8 moist–4/6 dry). A thin hard pan of iron compounds is frequently present in the middle reaches of this bed. The underlying coarse sand bed is always reddish, predominantly of the 5 YR hue.

Butzer and Hansen (1968) give a short description of the undulating Darb el-Gallaba gravel plain and appear to be of the opinion that the plain originated in late Pliocene time as a lacustrine basin fed by a southern river that passed through Wadi Kubaniya and the modern valley. These authors correlate the Wadi Gallaba gravels with the 150–160 m high gravels in the Kom Ombo graben across the river and consequently conclude that both were related, in early Pleistocene time, to one drainage system that derived its waters from the wadis of the Eastern Desert. It is now apparent that the gravels on the two sides of the Nile are different with regard to their mineralogy; the western Idfu gravels are of almost exclusively made up of quartz, while the eastern Kom Ombo gravels are of polygenetic origin including a large variety of igneous and metamorphic pebbles.

The Idfu gravels are noted in many parts of Nubia and Egypt. Said and Issawi (1964) describe terraces of an "earlier cycle" of the Nile sediments in Nubia which may be coeval with the Protonile. These are described as made up of "quartzite gravel embedded in a matrix of red-brown muds." They have an elevation of 32 and 21 m above the high water reservoir level prior to the erection of the dam. Butzer and Hansen (1968) describe similar beds overlying the Nubia sandstone bedrock in Nubia. Although of no great areal extent, the gravel terraces are classified according to their altimetry into five terraces: Wadi Korosko (23–25 m), Dakka (30–35) Adindan (40–42 m), Dihmit (44–48 m), and Wadi Allaqi (50–55 m). All are described as made up of quartz gravel overlying bedrock and in thicknesses not exceeding 7 m. A deep red to yellowish red (2.5–5 YR 5/6) paleosol with a coarse subangular blocky structure overlies these gravels. The gravel matrix is described as "slightly clayey, coarse sand that has presumably been eluviated of fines." The Dihmit and Adindan gravels are probably the most correlatable with Idfu gravels. They occur in fluvial platforms and are made up of rounded to subrounded pebbles of quartz and ironstone.

From Nubia, Giegengack (1968) describes similar beds under the term "Early Nile gravel" where the typical representatives of this unit include cobbles 10-20 cm in diameter and a finer grained sediment that fills the interstices in the gravel. The cobbles are mostly "pegmatite quartz, metamorphic quartzite, other crystalline lithologies, and a variable proportion of chert". In addition, large cobbles and boulders of Nubia sandstone are observed at the base of these gravels. The fine-grained fraction of Giegengack's Early Nile gravel consists of quartz grains coated and imperfectly cemented together by iron oxide in the form of hematite. After passing through a 25-cm thick superficial layer of surface gravels which is drab in color (5 YR 8/4), the section passes through a cemented conglomerate layer of deep red brick color (2.5 YR 4/4) 2.75 m thick. This is followed underneath by a loose gravel layer whose top is pallid in color (5 YR 8/4) which seems to mark the depth to which intense oxidizing conditions prevailed either during or after the concentration of $Fe_2 O_3$ in the section. Giegengack (1968) gives a map of the distribution of these gravels in Nubia (reproduced in figure 38) which are seldom observed further than 1 km from the present Nile. Patches of these gravels appear parallel to the Nile at Adindan east, Abu Simbel east, Aniba west, Korosko east, Wadi el-Sibu east, Seiyala east, Allaqi east, and Aswan east.

Recent drilling in the elevated arch which separates Wadi Tushka from the great oasis depressions of the Western Desert of Egypt shows an 11-m thick section of Idfu gravels overlying the Nubia sandstone (Figure 32), the base of which is at the same elevation of other Protonile sediments. This most interesting discovery could be evidence that the Protonile flowed into the great Kharga-Dakhla depression which is separated from the Nile Valley by an arch only 60 m higher than the present Nile Valley. This arch formed a great northwest-southeast horst crossing the Nile across the Allaqi depression (see Figure 47). Some authors advocate the idea that the oasis depression was flooded by the Nile during its recent history. The hypothesis, however, has to be excluded because no alluvial deposits or beach features are found or recorded in the depression, although the mineral composition of the Kharga and Dakhla soils is remarkably similar to that of the Nile (Anwar and Khadr, 1958). Further work is needed to attest the extent of the flooding of the Protonile into the oasis depression in the light of this interesting discovery.

To the north of Darb el-Gallaba plain, where the Idfu River seems to have branched into numerous braided channels, the path of the river is less defined. Patches of the Idfu formation are seen along the western bank of the Nile as far north as Esna to reappear again at the latitude of Minia where they continue in a more or less defined channel about 10–15 km to the west of the modern Nile until Cairo. Here the river seems to have negotiated its way through the Abu Roash outlier and continued to the north forming a more or less continuous channel, the relics of which are seen along a path that is 10 to 15 km to the west of the Rosetta branch of the modern delta of the Nile. The Idfu formation in this northern reach coincides with Sandford and Arkell's (1929, 1939) Plio-Pleistocene gravel terraces. In the middle latitudes of Egypt relics of the old channel of the Idfu River may be sought in the gravel terraces to the west of Minia, which appear on the older Geological Map of Egypt (1928) under "gravel spreads of uncertain age." These occur again as spreads of gravel running parallel to the Nile and resting directly over Eocene bedrock. Drilling in search of suitable limestones for the iron and steel industry in Egypt in this reach by the Geological Survey shows that these fluviatile gravels (composed mainly of rounded flint pebbles and cobbles) have a thickness of 30 m. Otherwise the channel seems to have been exhumed and buried under more recent cover of dune sand and local outwash. Further geophysical work could delimit this old channel if remains of it are still extant. In all these occurrences the Protonile assumes high elevations in comparison to the modern river (90–120 m above the modern flood plains).

In the subsurface there seems to be no record of deposits which belong to the Protonile; the

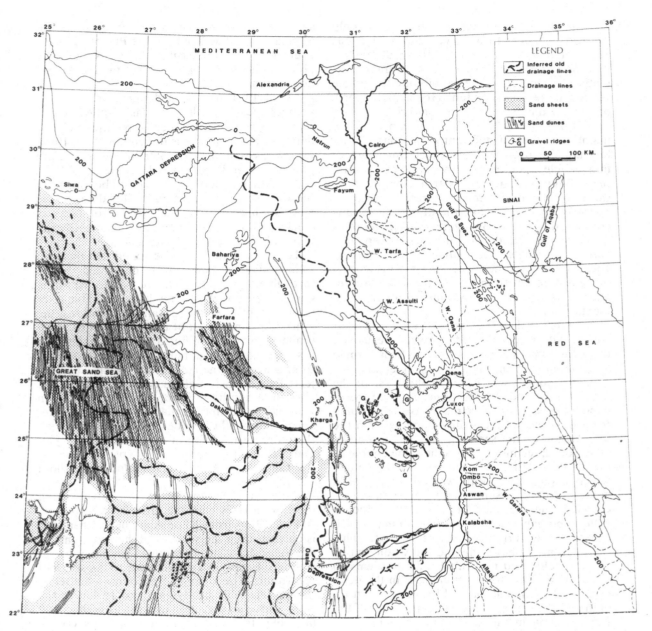

Figure 32. Map of Egypt showing possible drainage lines in Neogene and Quaternary times.

late Pliocene sediments underlie unconformably the graded sand–gravel unit of the succeeding Prenile. In certain wells, however, the lower part of the graded sand–gravel unit may be separated by virtue of its inclusion of several layers of silt. The El-Wastani formation (type locality El-Wastani well#1, lat. 31° 24, 8" N; long. 31° 35, 46" E) is a distinct unit which underlies the Prenile sediments. The thickness of this unit in this well is 123 m and occurs between depths 1009 and 886 m from the surface. Some brackish water foraminifera are recorded from a few of the clay beds that intercalate this formation. The unit shows large forset beds due to progradation and deposition in a fluvial to deltaic environment. The unit underlies unconformably the graded sand–gravel unit. It could well form part of the Paleonile system.

The Idfu gravels differ conspicuously from those of the Armant formation as they contain pebbles which are not derived from a local source but are transported from a distant source and are conspicuously smaller than the boulders of the Armant formation. The pebbles are mainly of flint derived from a deeply leached terrain with a small influx of igneous and metamorphic material; this characterizes this formation from earlier or later gravel formations. Good exposures exhibiting the whole section are few, but at Mena House to the north of the Pyramids of Gizeh, Shukri and Azer (1952) describe a 26-m thick section made up mainly of sands of different color shades with few intercalating beds of gravels made up of flint pebbles. The mineral analysis of 13 samples of this section shows that close to two-thirds of the heavy mineral fraction is composed of iron ores, while the other one-third is composed of the following minerals in decreasing order of abundance: amphiboles, epidote, garnet, staurolite, tourmaline, zircon, pyroxenes, and other minor accessories. This assemblage is different from that of the modern river and points to a derivation from a southern sedimentary terrain.

In Nubia and in the entire stretch of Upper and Middle Egypt the Idfu gravels are mostly deposited on a bedrock surface which formed the bottom of a through-flowing river that occupied a course which lay to the west of the modern Nile and the Paleonile. At the time of deposition such a surface must have constituted the lowest part of the valley. Judging from the distribution of the sediments of the Paleonile, which occupied more or less the same course as the modern Nile, and that of the Protonile, which occupied a more westerly channel, it would seem that some tilting must have occurred along the axial course of the modern valley, perhaps during Issawia time, which caused the lowering of its western part. This tilting seems to have affected the whole valley and the delta; and the Idfu Protonile seems to have had a single westerly channel in the delta region. This channel passed by the eastern tip of the Wadi Natrun depression. The path of the Idfu Protonile in the Faiyum region seems to indicate that the Protonile did not have access to the Faiyum depression.

The height of the present Idfu Protonile sediments indicates a gradient that might have been even smaller than that of the modern Nile. In Nubia and in the southern reaches of Upper Egypt the gravels are in the range of 60–80 m higher than the modern flood plain, while to the north the gravels are 80–100 m higher than the modern flood plain. Because these elevations do not represent the actual surfaces of deposition, which were the sites of erosion and lowering since their elevation, it is difficult to come to

any conclusion regarding the gradient of the original river. The height of the Protonile, however, would make possible the breaking of the Nile across the Tushka arch into the great oasis depression.

The Idfu Protonile sediments seem to be archeologically sterile. Records of rolled "Chellean" artifacts are frequently mentioned. The dating of the Protonile is difficult. Hayes (1964, p. 26) gives a date of 660,000 years before present to Sandford and Arkell's 320 foot "terrace," which is here considered as belonging to the Idfu Protonile sediments. This date fits the stratigraphic position of the Protonile as worked out in this work, and would classify the Protonile as of early Pleistocene age.

To recapitulate, it can be stated that the Idfu sand–gravel complex was deposited by a through flowing river capable of transporting cobble-size gravel for long distances. This river seems to have had the same sources as the Paleonile. Since most of the gravels of the Protonile are made up of flint and quartz, it must be assumed that these must have been derived from a deeply leached terrain which was subjected to intensive chemical weathering. The climate of southern Egypt seems to have been wet and somewhat similar to that which prevailed during the Pliocene. The Idfu pluvial, therefore, was different from all other "pluvials" of the Pleistocene in Egypt inasmuch as the rains were evenly distributed throughout the year.

If my earlier contention of the equivalence of the Idfu gravels with the gravels described by Arkell (1949) from Khor Anga near Khartoum is correct, then the sources of the sediments of the river have to be sought, in addition to southern Egypt, in a provenance to the south of Khartoum. The river had a course parallel to the modern Nile but lying 10 to 15 km to the west, and at an elevation that ran as high as 100 m above the modern flood plain. Its course followed that of lower Nubia and Egypt.

Judging from the volume of sediment deposited by this river, which is estimated not to exceed 300 km^3, the river seems to have been of short duration, centered probably around the 700,000 B.P. episode of intense climate changes in the northern hemisphere (Emiliani 1966, Hays et al., 1969; Kukla, 1977). The river did not seem to have fanned out into a delta in the land of Egypt but seems to have flowed to the sea in a single, though braided, channel. It is difficult at present to know whether this river continued northward in the elevated Rosetta fan of the Mediterranean basin or not. No sediments which could be compared to the Idfu Protonile are described from the boreholes of the North Delta embayment. The Wastani formation lies below the sand–gravel beds and includes a large number of clay intercalations. Furthermore, the Protonile ran in a course which lay at an elevation of 80–100 m above the modern flood plain.

The Idfu Protonile sediments are characterized by a mineral assemblage which is different from that of the modern Nile. The gravels are well-sorted, rolled, and composed mainly of flint and quartz with minor ingredients of igneous and metamorphic pebbles. In this respect, the Idfu gravels differ from the earlier and later gravels in the Nile section. The gravels of the Idfu Protonile have a red soil which is probably the most distinctive characteristic of the sediment. Along the Darb el-Gallaba plain, the Protonile gravels of the Idfu formation are now seen capping the high bedrock outcrops of the Nubia sandstone that interrupt the monotomy of the plain, indicating that the formation had a greater thickness and that its deposition was followed by a period of intense erosion.

II-6. THE PRENILE (Q₂)

The Prenile represented a vigorous and competent river with a copious supply of water and a wide flood plain. Its sediments are coarse, massive and thick. Its sources seem to have been outside Egypt and there is evidence that the Ethiopian highlands were important in the supply of sediments. This river flowed for a long period of time covering most of the middle Pleistocene and terminating around 200,000 B.P. late Acheulian implements occur in the top layers of the Prenile sediments and in large number in the overlying deposits. During this long period great changes took place in the Nile Valley and in the lands from which it derived its waters.

Fluviatile sediments belonging to this complex river system crop out in a most conspicuous way along the banks of the Egyptian Nile Valley and the delta margins forming an important element in the landscape of the valley. They seem also to fill the deeper parts of the channel and flood plain of the modern Nile Valley and delta where they appear consistently on the logs of all wells drilled in these reaches. The graded sand–gravel unit belonging to this formation lies unconformably beneath the famed agricultural silt layer of the fertile land of Egypt.

The Prenile sediments represent a thick and easily recognizable unit made up mainly of sands which seem to have been deposited by a competent river. These sediments had been previously divided into the Qena and Dandara formations (Said *et al.*, 1970) and had been considered the deposits of a through-flowing river, the Prenile. Recent field work, however, has made possible the rectification of the position of the Dandara formation within the stratigraphic column of the Nile sediments. It has now been shown, contrary to the author's earlier contention, that the Dandara overlies the Qena formation unconformably and does not seem to belong to the same river system. Because of its fine-grained lithology, texture and general aspect the Dandara may best be classified with the sediments of the less vigorous succeeding river marking the advent of the tenuous transitional episode in which the Neonile established its regimen after long intervals of little flow.

The Qena sands form an exceedingly uniform and characteristic unit which skirts the eastern and western banks of the river in Upper Egypt. They have a characteristic topography. They are always overlaid by the Abbassia gravels. Along their slopes accumulated the Makhadma wash of pre-Neonile age indicating that the rolling topography of the Qena sands was formed during the Prenile–Neonile interval, and that since then it had undergone little change.

The Prenile Qena sands are recognized by earlier authors. Sandford and Arkell (1929, 1933, 1939) and Sandford (1934) describe these sands and give them an age ranging from the "Pliocene" to the "Lower Paleolithic." Butzer and Hansen (1968) follow the practice of these authors and ascribe the sands of the Kom Ombo area to the Pliocene. The type Qena formation is described by Sandford and Arkell as part of the Pliocene Gulf deposits interfingering and interdigitating the marl–conglomerate materials of what is considered in the present work as the Armant formation and the masses of limestone breccia of what is here called the Issawia formation. This is advocated in spite of the fact that Sandford and Arkell (1933, p. 13) admit that the sands of Qena had a different source from those of the other locally derived formations, for they were "brought from the south by the main stream which fed the headwaters of the Gulf above Kom Ombo." Our observations show that the Qena sands form a unit

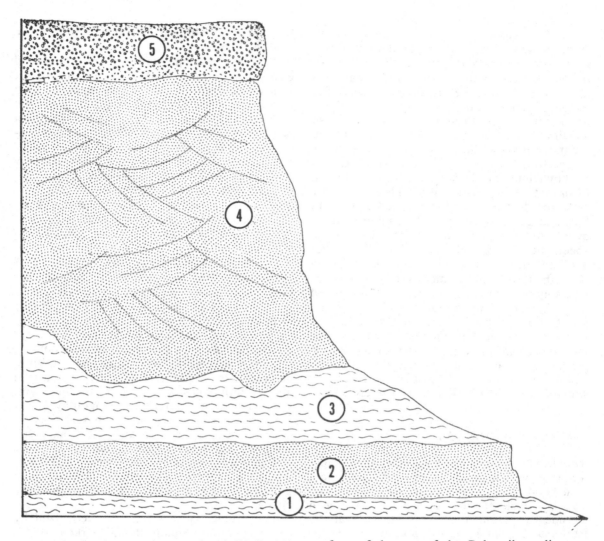

Figure 33. Section at the mouth of Wadi Abu Nafukh, opposite Balyana, Upper Egypt. Reinterpreted after Sandford (1934).

Abbassia formation
5. Gravel, 5 m thick
Prenile Qena sands
4. Sandstone cross-bedded, 32 m thick, resting on eroded surface of bed 3
Paleonile Madamud formation
3. Marl, brown
2. Quartzose sandstone, white
1. Marl, buff with brown specks

which lies above the units mentioned by Sandford and Arkell and, in fact, is separated from them by a marked unconformity and a long period of time. The Qena sands represent the deposits of the middle Pleistocene.

Recent field work has also uncovered the base of the Qena sands. At the mouth of Wadi Abu Nafukh (Figure 33) they rest on the eroded surface of the top of the Paleonile sediments made up of a series of marls and fine-grained quartzose sandstone beds. The Qena sands assume 32 m in thickness in this locality. They are made up of a massive cross-bedded quartzose sandstone unit with specks of feldspar grains. They underlie a unit of gravel about 5 m thick and include pebbles derived from the red breccia of the lower Pleistocene Issawia formation.

The Qena sands assume different geomorphological aspects along the banks of the Nile. In the south they form hills with rounded tops which constitute an important element of the rolled topography of the Qena-Kom Ombo reach. In the north they form benches of great areal extent. Although these sand formations differ in geomorphology they seem to belong to one natural system. Both facies have a similar mineral composition and carry a molluscan fauna of African aspect. It is possible, however, that these units may not be contemporaneous, but they certainly belong to one hydrological regimen.

The mineral composition of the Prenile sediments differs from that of the older Protonile sediments by the presence of pyroxenes in appreciable quantities. In spite of the fact that the varieties of the pyroxene minerals found in the Prenile sands are similar to those of the modern tributaries of the Nile in Ethiopia, the Prenile sands differ from the modern sediments of the Nile in having smaller amounts of epidote (see also Hassan, 1976). Granting the difficulty of arriving to conclusions with regard to the provenance of the sediments on the basis of a small number of mineralogical analyses, it is nonetheless feasible to believe that the Ethiopian highlands contributed sediments to the Prenile.

The sediments of the Prenile carry an African molluscan fauna which is different in the southern facies from that in the northern. Said *et al.* (1970) report the presence of arenaceous casts of *Unio abyssinicus* and *Aspatharia cailliaudi* in the Qena sands of the Dandara area. In the north shells of *Corbicula artinii*, *Unio gaillardoti*, and *Mutelina aegyptiaca* are ubiquitous in the Qena sands. This difference in faunal assemblages may point to a younger age for the northern facies, although all species recorded from both sands are still extant. This adds credence to the earlier contention derived from the study of the mineralogy of the Prenile sediments that, for the first time in the history of the river, a sizable portion of its waters came from the Ethiopian highlands.

The type section of the Qena sands is Wadi Abu Mana'a quarry face (Figure 34) which exhibits a magnificent section of these cross-bedded flood-plain deposits of an extremely competent river. The section is made up of a massive 22-m thick bed of sand with very few bands of conglomerate whose pebbles are well rounded and made up of a variety of basement rocks. Mechanical analyses of five samples collected along the face of this quarry show these sands to be well sorted; more than 60% of the sediment by weight falls within the 630–200 μ size range. Only in one sample is the bulk in the more than 1000 μ size range. The clay fraction of the samples never exceeds 8% by weight of the sample. The mineral analysis shows that the heavy minerals constitute about 2% of the

Figure 34. Abu Mana'a quarry face.

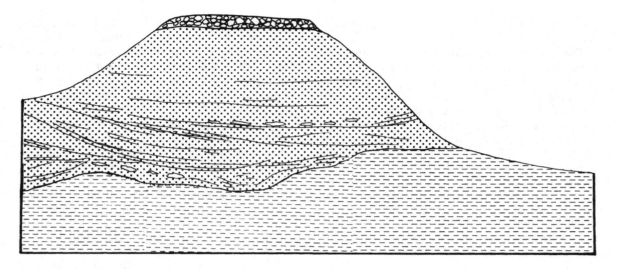

Figure 35. Section of the cliff on the west bank of the Nile opposite Aklit, Upper Egypt. After Sandford and Arkell (1933).
  Abbassia formation
  4. Conglomerate, polygenetic, overlying unconformably bed 3.
  Prenile Qena formation
  3. Coarse quartzose sand, nonfossiliferous
  2. Laminated micaceous sand with concretions of clay
  Cretaceous bedrock
  1. Dakhla shale

sand fraction of these samples. The opaque minerals make about 50% of the heavy minerals present. The other 50% is made up of the following minerals in decreasing order of abundance (average of five samples): epidotes (25%), pyroxenes (12%), garnet (6%), tourmaline, zircon, staurolite, rutile, and other minor accessories (7%).

A number of exposures were mapped and measured on both banks of the Nile in the Kom Ombo area. Along the west bank opposite Aklit, Sandford and Arkell (1933, pp. 8–10) describe a section made up of cross-bedded micaceous sands and laminated clays, passing up vertically into coarse quartz sands. The uppermost sands are set in a white calcareous matrix containing root drip and other organic impressions and nowhere excceeding 130 m in elevation. The sands (Figure 35) overlie disconformably the Cretaceous Dakhla shales and underlie unconformably a cemented polygenetic gravel bed, the Abbassia formation (see below). Butzer and Hansen (1968) take these sand deposits, now assigned to the middle Pleistocene Prenile, as representing the sandy Nile facies of the Pliocene in Egypt.

The Qena sands make up the Burg el-Makhazin hill in the Kom Ombo plain (the hill was previously mapped as a bedrock of Nubia sandstone in the Geological Map of Egypt, 1928, 1971). The 16-m thick micaceous sands of this hill are loose, erratically bedded, separated by thin seams of clay and are graded from coarse to fine sand. They are quarried, like many other Qena sand exposures, on a large scale. The following is a section measured in the eastern Kom Ombo plain (Figure 36):

Abbassia gravels
  • Conglomerate made up of pebbles of igneous rocks, quartzite and siliceous sandstones; size varying from 2 to 20 cm in diameter; all embedded in a matrix of reddish-brown soil; thickness 8.4 m. The mechanical analysis of this conglomerate bed shows that 73% of the pebbles are coarser than 5 cm in diameter.

Qena formation
  • Silt, yellowish, containing nodules of quartzitic sandstone, forms a wall; thickness 6 meters. The mechanical analysis of one sample of this silt shows that close to 96% of the sample is in the silt–clay fraction.
  • Sand, loose, coarse becoming medium-grained toward the upper part of the bed, yellowish to grayish white, cross-beeded, containing brown clay nodules, abundant root drip; forms a slope; thickness 13.8 m. The mechanical analysis of a sample of this unit shows that 40% of the sample is coarser than 1 mm.

The mineralogy of four samples from the Qena sands of this section is more or less uniform. The opaques range from 45 to 84% with an

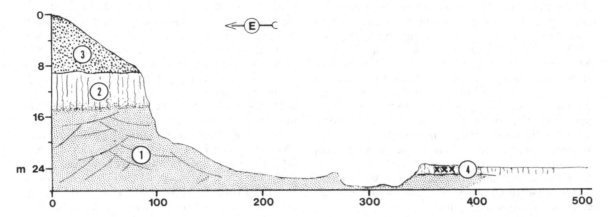

Figure 36. Section at Burg el-Makhazin, Kom Ombo plain, Upper Egypt.

Neonile sediments

4. Silts carrying archeological materials

Abbassia formation

3. Conglomerate, polygenetic resting unconformably over 2

Prenile Qena sediments

2. Silt grading into

1. Sand, loose, cross-bedded

average of 63%; the epidotes range from 12 to 39% with an average of 21%, and the pyroxenes range from 5 to 10% with an average of 7%. Amphiboles, garnet, zircon, tourmaline, and other accessories make up for the remainder of the heavy mineral fraction.

In the north, the Qena sands fringe the western part of the cultivation edge all along the Nile in Middle Egypt. The most continuous stretch of the Prenile sediments is that which extends all along the western bank of the Nile from Manfalut to Wasta and from there to Cairo. Seen from the valley it forms a continuous bluff, usually incised by small gullies. In this stretch sand dunes interfinger these deposits in a manner which is very similar to that occurring within the Neonile sediments (Wendorf and Said, 1967). This seems to indicate that similar climatic and physical conditions must have prevailed over Egypt during the deposition of the Prenile sediments as prevailed over Egypt during most of the Holocene. The Prenile sediments in this stretch are described by Sandford and Arkell as forming their 44-m terrace.

The mineral analysis of the sands of this terrace in a locality opposite Manfalut shows that it is similar to that of the Qena sands of the southern reaches of the valley (see also Shukri and Azer, 1952, where the analysis of the 44-m terrace of the Faiyum stretch is given). In this

"terrace" also about 50%,of the heavy minerals separated belong to the opaques. Amphiboles come next in abundance and then epidotes, pyroxenes, garnet, and other accessories. The mineral analysis of all sand exposures along the Nile Valley is, therefore, the same except for the presence of amphiboles in the seemingly younger sands of the north. It seems certain that this is not due to provenance but to the size of the fraction from which the heavy minerals were separated. The fact that two sand units have more or less the same mineral composition, in addition to their similar lithogical appearance, gives support to the idea that both form a natural group that belongs to one, though complex, hydrological system.

The Prenile deposits outcrop along the margins of the delta on both sides of Wadi Tumeilat, which forms the southern fringe of the lands to the east of the modern delta, and seem to extend across the Suez Canal into Eastern Sinai. On the western margins of the delta the Qena sands appear again as low terrace-like extensions covering the desert fringes. To the northwest of Cairo, the Qena sands fan out along the Cairo-Alexandria desert road forming the characteristic rolling topography of this stretch. Many of the hills in this area were opened up and extensively quarried. Magnificent exposures of these massive, uniform, cross-bedded flood-plain loose sands can now be seen in this area. Along the eastern edge of Wadi Natrun similar exposures were also opened up. These represent the most northerly exposures of the Qena sands. To the north of Wadi Natrun, the Qena sands disappear under subrecent stabilized dunes which are known to fringe the desert on the edge of the Beheira province.

In the Mediterranean basin the Nile cone seems to have been built to a large extent by the Prenile sediments. Ryan *et al.* (1973b) studied the heavy mineral composition of the sand layers in the piston cores from the western Nile cone (Rosetta) collected during the Deep Sea Drilling Project Cruise Leg 13 in the Eastern Mediterranean. They find that while the surface samples have a mineral composition similar to that of the modern Nile, the drill core samples, which are "of an appreciably older Pleistocene age than the surface piston cores," have a higher epidote content. The samples thus described from the core of the Rosetta cone seem to belong to the Prenile sediments. This seems to indicate that the Prenile must have reached as far seaward as the Strabo trench of the Eastern Mediterranean basin.

Subsurface Prenile sands and gravels of very similar lithology to those cropping out are found in all boreholes drilled in the valley and delta of the Nile beneath the agricultural silt layer of the fertile land of Egypt. They represent the water-bearing horizons in the valley and the delta. These are correlated on lithological grounds with the Prenile sediments.

In the delta wells a consistent unit of thick layers of quartzose sands and pebbles occurs above the el-Wastani formation. It is given the name Mit Ghamr formation. The type locality is at Mit Ghamr well #1 (lat 30°41'44" N; long. 31°16'26"E) between depths 20 and 483 m. The sands are medium to coarse-grained. The pebbles are mainly quartz, but occasional silicified limestones and chert pebbles are also found. In places interbeddings of coquina of marine shells occur. Some peat layers are also observed. The upper levels of this formation are marked by the appearance of thin beds of clay, silt, and peat containing coastal or lagoonal faunas. These riverine sands appear also in outcrop as isolated low mounds or "islands" in the midst of the agricultural fields of the delta of the Nile representing the higher parts of the eroded surface of this complex. They are the "turtle backs" described by Sandford and Arkell (1939), the most famous of which is that of Quesna to the north of Cairo. The mineralogy of the sands of this turtle back is given by Kholeif *et al.* (1969) where it is found to be similar if not identical with that of the Prenile Qena sands.

The thickness of the Prenile sediments is in the range of 250 m in the valley. In the north of the delta, the thickness may exceed 1000 m. Figure 37 gives the isopachs of the Prenile sediments which show that the thickness is greater in the axial parts of the modern delta and increases as one proceeds toward the Mediterranean Sea. Thus, in the middle latitudes of the delta thicknesses of 600 m are common, while

near Cairo the thickness decreases to about 250 m. Many boreholes dug outside the recent flood plain of the river in the fringes of the delta show that the deposits of the Prenile extend for long distances outside the reaches of the modern delta. The maximum thickness of the Prenile sediments lies to the west of the modern channel, and it is in this western channel that the main stream seems to have run.

A glance at the accompanying geological map shows that the Prenile occupied a course which lay to the west of the modern Nile and to the east of that which was occupied by its predecessor, the Protonile. This may be related to some tilting of the land to the west of the Nile. The Prenile sediments appear also quite conspicuously in the eastern reaches of the delta; this may be explained as due to an activation of the Pelusium line which brought these reaches within access of the Prenile. This line of faults lies along one of the modern seismic zones of Egypt (Gergawi and Khashab, 1968a) and is considered as one of the main structural lines of Egypt by Neev (1975) and Said (1979).

It seems also that at some time in its history, the Prenile broke through the Faiyum depression; terraces of young Prenile sediments, indistinguishable from those of the Nile-Faiyum divide, encircle the depression in many places at an elevation of 44 to 46 m above sea level (Little, 1935). Previously, Said *et al.* (1972a,b) had noted the deposits of this beach but had expressed doubts as to its lacustrine character and association with the Faiyum depression. Our field work has now established the fact, already noted by Little, that the Prenile in its late phases entered the Faiyum depression, and relics of its deposits are now seen encircling it at a constant elevation on all sides especially along its southwestern rim (Figure 51).

In conclusion, it can be stated that the Prenile sediments are uniform. They include the fluviatile sediments of a through-flowing competent river which derived a large part of its waters from the Ethiopian highlands. The Prenile sediments cover large stretches of the Nile Valley and the fringes of the delta and form an important element of the landscape of the valley and delta.

The sediments of this stream reach great thicknesses, about 250 m in Upper Egypt of which about 200 m are known from the subsurface of the modern valley beneath the agricultural layer deposited by the succeeding and extant Neonile. In the delta region the thickness of the Prenile sediments exceeds 1000 m. It can be safely said that the Prenile was the largest

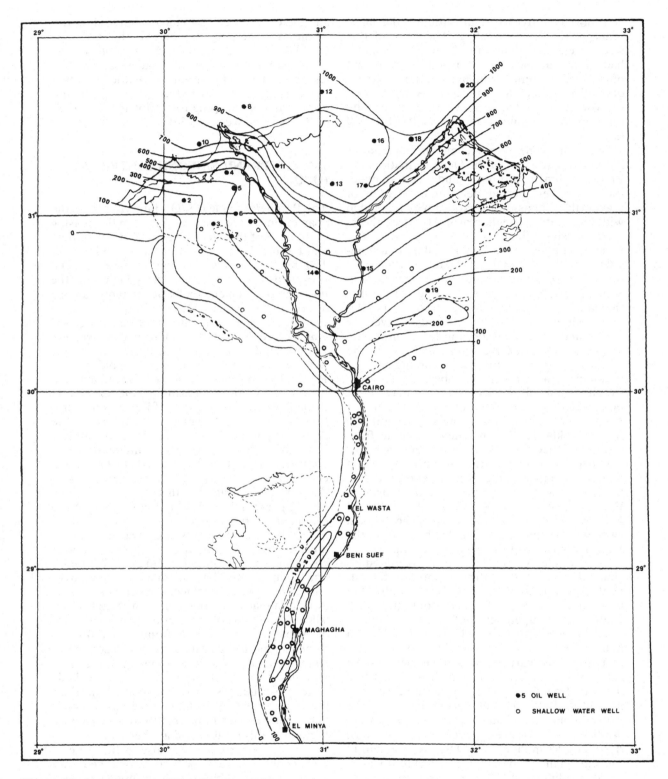

Figure 37. Isopach map of Prenile (Q₂) sediments. Numbers refer to wells, the locations and stratigraphic data of which are given in Table A-1.

and most effective river in outlining the landscape of the modern valley. The delta of this river extended well into the Mediterranean and had an area at least three times as large as that of the modern delta. Judging from the thickness of the Prenile sediments it can be estimated that the volume of sediment that was deposited by this complex Prenile system exceeded 100,000 km³.

Save for some late Acheulian implements found in the top part of the Qena formation, the Prenile sediments do not include datable materials. However, the sediments underlie a unit of gravel (the Abbassia formation) which is laden with fresh late Acheulian artifacts. It is, therefore, possible that the Prenile system terminated at about 200,000 B.P., a date that coincides with that conjectured by Hayes (1965). Although there is no absolute date for the late Acheulian in Egypt, it is interesting to note that the date estimated here on the basis of its relative position coincides with the date given by Szabo and Butzer (1979) for lacustrine limestones carrying late Acheulian artifacts in the Kimberly district, South Africa, on the basis of uranium series dating. The beginning of the Prenile is difficult to ascertain but it may be set perhaps at 700,000 B.P., a date that marks a major event in the Pleistocene sequence in the world (Kukla, 1977). This may not be far from the truth since the Prenile represents a major event in the history of the Nile and Africa. It marks the dawn of the Nile as known today, linking it for the first time with its source in Ethiopia and most likely with the great Sudd Lake which it drained by crossing the Nubian swell which separates the Sudan basin from the Egyptian basin and the Mediterranean.

The Prenile system, therefore, excavated its channel to the east of the Protonile at the advent of the middle Pleistocene, shortly after the latter had ceased to flow. It continued with vigor and competence until late Acheulian times, a period that extended for about half a million years. The annual suspended matter that this river carried must have been in the range of 0.2 km³, a figure which is almost five times that carried by the modern Nile prior to its control. Ball (1939) points out that out of the 110 million tons of suspended matter passing Wadi Halfa in an average year by the modern river only 58 million tons (0.05 km³) remain in suspension in the river at its passage past Cairo, corresponding to 0.027 km³ of sediment compacted to 10% moisture content by dry weight.

Most of the waters of the Prenile system seem to have come from sources outside Egypt,

and the climate seems to have been arid on the whole. Wadi activity took place in the earlier intervals as judged by the presence of several conglomeratic beds in the Quna sands. The later episodes of the Prenile were arid as shown by the presence of wind-driven sands in the upper parts of the Qena formation in the stretch of the Nile between Assiut and Cairo.

II-7. THE PRENILE/NEONILE INTERVAL (Q₂/Q₃)

The Prenile/Neonile (Q₂/Q₃) interval covers the span of time which elapsed between the ceasing of the flow of the Prenile in late Acheulian time and the breaking through of the Neonile about 120,000 B.P. During this interval a great pluvial prevailed over Egypt, and rains falling on the uncovered basement rocks of the mountainous areas of the Eastern Desert brought down to the Mediterranean great quantities of gravel which were deposited unconformably over the Prenile sands. The new ephemeral river seems to have followed the channels of the older river. Red and yellow-red soils were formed, and man made a grand appearance in the Nile Valley and in many parts of the deserts of Egypt. The duration of this pluvial, which is here termed the Abbassia pluvial, is not known; but an estimate of 80,000 years (perhaps contemporaneous with the Riss/Wurm interglacial in the Alpine classification) is reasonably supported by dates that precede and succeed this pluvial. In the following paragraphs a description is given of the sediments of this pluvial.

The Prenile sediments are capped in many areas of the valley by a thick fluviatile gravel cover mostly of pebble size and diversified mineral composition. The thickness of this cover varies from place to place. It averages 6 m although in places the thickness may reach 15 m. The most famous of these gravel beds are those of Abbassia, near Cairo, from which Bovier-Lapierre, more than 50 years ago, described *in situ* implements of what was then described as a stratified site about 10 m thick. These beds were described as including eoliths in their lower part and Acheulian materials in their upper part. Bovier-Lapierre (1926) gives a photograph of this locality and a schematic description of the occurrence. Because of the fame of this locality these important gravel beds are given the name Abbassia, even though Bovier-Lapierre's site cannot be seen again, for it has been tilled down to form the foundations of the buildings of Nasr City, a new suburban area which is being developed near Cairo. I propose, therefore, to take as a type the Rus section situated at the mid-desert station of Rus along the state railway line from Wasta to Medinet el-Faiyum traversing the Prenile bed at its broadest part in the Nile-Faiyum divide. The Rus station is the center of extensive quarrying of the gravels for railroad ballast. The ballast pits, about 8 m thick, expose a massive gravel bed made up of well-rounded and well-sorted (mostly 5–10 cm in diameter) pebbles of red, green, and purple rocks derived from the basement rocks of the Eastern Desert of Egypt. The whole rests unconformably, as in Abbassia, over sands of the Prenile complex. Sandford and Arkell (1929) note that the archeological materials separated from the lower parts of this formation are water-worn, while those separated from the upper parts—rare examples of beautiful Acheulian work—are "almost as fresh and as sharp as on the day of their manufacture." These authors, therefore, describe the Abbassia "terrace" as the Acheulian terrace, a contention with which the present author is in full agreement. Bovier-Lapierre's Abbassia site must have been similar, and the cultural materials separated from it must be reexamined in the light of this new stratigraphic information.

The Abbassia gravels are widely distributed in the valley and delta regions of the Nile. They make some of the best quarrying sites for gravel in Egypt. They follow on top of the Prenile sediments, truncating them, and are separated by a clear-cut unconformity. In places, however, such as in the Cairo area, the lower part of the Abbassia section includes several thin bands of the Qena sands, indicating that the Abbassia pluvial was not separated by a long lapse of time from the Prenile riverine conditions; in fact, it probably started during the waning stages of the Prenile.

In Nubia Giegengack (1968) describes several inverted wadi beds which he names "Wadi Conglomerates". These gravel ridges are younger than Giegengack's Earlier Nile gravel (figure 38) and are regarded as early Wurm by Giegengack. The ridges seem to form a drainage system integrated with the Nile. The gravels are mostly of local derivation. Extensive search for archeological materials led to the discovery of a few hand axes of late Acheulian appearance. They are, therefore, correlated in this work with the Abbassia gravels. The dating of these inverted wadis as Acheulian shows the intensive depth of erosion to which Nubia and southern Egypt have been subjected to since the Acheulian. It is estimated that southern Egypt and Nubia were lowered by at least 50 m during and since the Abbassia pluvial.

The Abbassia gravels differ from the Idfu gravels in composition, the latter being made up of quartz pebbles derived from deeply leached terrain, while the former contain abundant crystalline rocks and feldspathic sands derived from a deeply disintegrated but little-leached terrain. The Abbassia gravels are widely spread in the valley and form characteristic beds in many parts of Egypt; they seem to include rich archeological materials. Of these one may mention Sandford and Arkell's (1939) 100-foot terrace at Sebaiya and east of Esna, the 50-foot terrace of the Kom Ombo plain and El Kab near Idfu and the 30-foot terrace near Armant. If the archeological materials found by Giegengack (1968) in the "Wadi conglomerates" in the inverted wadis of Nubia are authentic Acheulian, then these gravels could well belong to the Abbassia formation. All these "terraces" are characterized by similar lithology and archeological materials, and seem to belong to one formational unit which assumes these different elevations due to later erosion. The futility of using elevations in correlation is obvious in a valley with as complex a history as that of the Nile Valley having been formed by several streams following one another, obstructed by waterfalls, and separated by periods of intense tectonics and erosion. The mechanical analyses of the Sebaiya and Kom Ombo gravels show that about 30% of their pebbles are made up of quartz, 50% of basement materials (granodiorite, diorite, gabboro, syenite), and 20% of fine-grained materials, mainly sand and clay.

The Abbassia gravels owe their origin to deposition in a stream which had its headwater in the Egyptian deserts which must have enjoyed then a much wetter climate to justify calling this episode the Abbassia pluvial. The age of this pluvial is late Acheulian. There is archeological evidence that this was a period when main lived over most of the desert areas of Egypt as well as in the valley, and that there was a savanna type of vegetation and a roving Ethiopian megafauna in many of the now barren areas of these regions (Wendorf et al., 1977). The deposition of gravel beds derived from the Egyptian basement to the east of the Nile suggests an enormous wadi activity that has seldom since been witnessed.

The effect of this pluvial can be seen also in the number of soils that are preserved on older surfaces such as the red-brick to yellow-red soils of the Idfu surface and the red-brown soils of the 10-5 YR hue of the Abbassia surface.

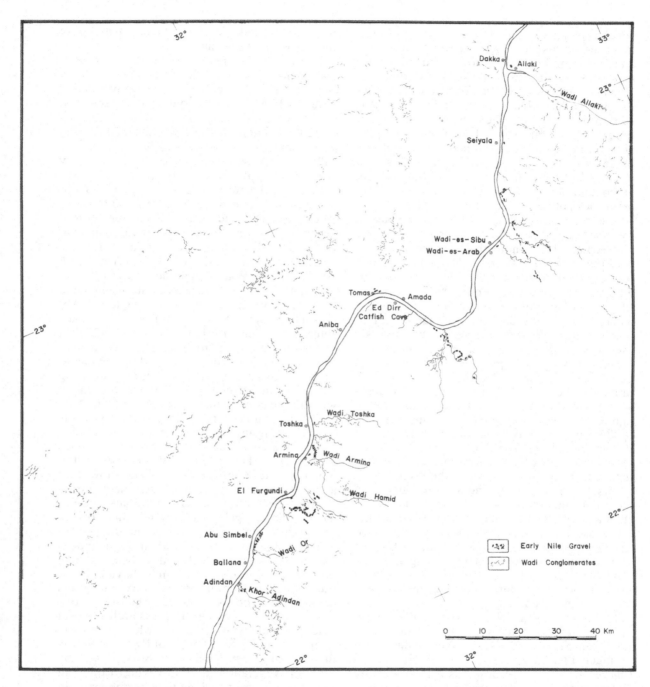

Figure 38. Map showing distribution of nilotic gravels (Early Nile and Wadi conglomerates). After Giegengack (1968).

The amount of rain that would be required to develop soils of this nature is discussed by Flint (1959), Giegengack (1968), Butzer (1964), and many others. According to most authorities red soils form in climates having an annual precipitation in excess of 40 inches and a semitropical to tropical mean annual temperature. Giegengack cites Walker whose work in Baja, California, shows that red iron oxide pigment in Pleistocene soils can develop as a result of diagenetic alteration of ferromagnesian minerals, chiefly hornblende, under conditions where the mean annual precipitation exceeds 6 inches. Giegengack compares the mineralogy of both soils to show that the Idfu soils display characteristics significantly different from those of Baja, California. While in the California soil, $CaCo_3$ is present in abundance throughout the soil profile, and the little clay that may be found is in the form of montmorillonite, the Nubian soil, according to Geigengack, has little $CaCo_3$ and abundant clay. However, in Egypt the soil which developed over the Dandara formation in its type locality shows an abundance of $CaCo_3$ underneath the soil, iron minerals are abundant, and the significant clay fraction is montmorillonite.

A typical soil profile over the Idfu gravels in its type locality in the Darb el-Gallaba plain is as follows:

0–50 cm	Coarse gravel, with slightly silty coarse sand between the gravel.
50–150 cm	Yellowish-red cemented coarse sandy silt (5 YR 4/8 moist–4/6 dry) with a thin hard pan of iron compounds at about 70 cm depth; very rich in lime
150–180 cm	Strong brown, medium fine sand cemented with soft lime aggregates (7.5 YR 5.8 moist–5.5/8 dry).
180–200 cm	Gray silt to clay with soft and hard lime aggregates (10 YR 6/1 moist–6.5/1 dry).

It thus seems that the Abbassia pluvial was a period in which extensive rains fell over Egypt to produce a through-flowing river which deposited thick polygenetic gravels, derived mostly from the uncovered basement rocks of the Eastern Desert, in channels that followed the modern wadis and in a master stream that closely followed that of the Prenile. The Nubian Desert seems also to have been drained by wadis into the Nile. Soils developed over old surfaces; they are mostly in the 5 YR hue in the younger formations and in the $2^1/_2$ YR hue in the older formations.

II-8. THE NEONILE (Q_3)

The deposits of the Neonile are made up of silts and clays indistinguishable in aspect and composition from those which were deposited over the land of Egypt by the modern Nile up to the very recent past. They rest over the eroded surface of the Prenile sediments with a marked un-conformity. These deposits form the top layer of the flood plain of the modern Nile and are also found outside this plain in the form of benches that fringe the valley at elevations ranging from 1 to 12 m above the modern flood plain. These sediments seem to have been deposited by a river which could not have been very much different in regimen and sources from the modern river.

The Neonile sediments were formed in four aggradational episodes during which structured silts were deposited. These episodes were separated by recessions during which wadi deposits, slope wash, pond or playa sediment, and/or diatomites accumulated. Shortly after the Neonile started flowing into Egypt about 120,000 years ago, a long recession took place during which a major pluvial and enormous geomorphological changes occurred leading to the shaping of the modern landscape of the valley. The succeeding aggradations were punctuated by relatively short recessional intervals. They may well form one series with the exception of the recession which marked the advent of the Holocene during which the valley was incised in response to a change in base level.

The last three aggradations of the Neonile have been recently subjected to detailed studies as a result of the international campaign to salvage the archeological treasures of Nubia. Detailed studies of the stratigraphy of these sediments in Nubia are given by de Heinzelin (1968) and Butzer and Hansen (1965, 1968); and in Egypt by Wendorf and Said (1967), Said et al. (1970, 1972a,b), and Albritton (1968). A review of the work carried out in both Nubia and Egypt is given by Wendorf and Schild (1976).

The aggradational episodes of the Neonile may, therefore, be broadly divided into three groups—the basal, older, and younger; the first two belong to the late Pleistocene and the last to the Holocene.

Younger Neonile deposits
  δ aggradational silts and fluviatile sands: Arkin formation
  γ/δ recessional deposits: Dishna-Ineiba formations

Older Neonile deposits
  γ aggradational silts and fluviatile sands: Sahaba-Darau formation
  β/γ recessional deposits: Deir el-Fakhuri formation
  β aggradational silts and fluviatile sands: Masmas-Ballana formation

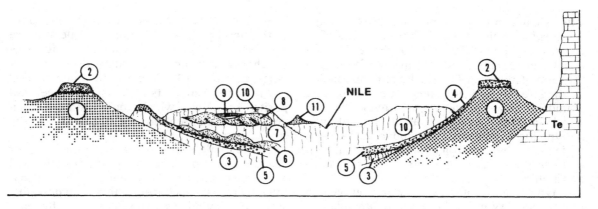

Figure 39. Hypothetical schematic profile of Nile Valley in Upper Egypt showing relationships of different rock units of Nile sediments.
11. Modern flood-plain and overlying dunes
10. Sahaba silts and fluviatile sands
9. Deir el-Fakhuri deposits
8. Ballana dunes
7. Masmas silts
6. Ikhtiariya dunes
5. Korosko wadi sediments
4. Makhadma slope-wash
3. Dandara silts
2. Abbassia gravels
1. Qena sands

Basal Neonile deposits
α/β   recessional   deposits: Korosko-Makhadma formations; Gerza-Ikhtiariya formations
α aggradational silts and fluviatile sands: Dandara silts

Figure 39 is a hypothetical schematic cross section in the Nile Valley showing the relationships of the Neonile and the older nilotic sediments; Figure 40 is a schematic diagram of the older and younger aggradational sediments of the Neonile showing the disposition and relationships of the different units to one another.

The Basal Neonile Deposits
The Basal Neonile deposits are made up of an assortment of sediments including the earliest aggradational silts of the Nile with modern aspect; the overlying dunes and soils of the Gerza arid episode and the slope wash and wadi sediments of the Mousterian-Aterian pluvial. In the following paragraphs a description of the deposits of this Basal group is given under the headings: Dandara, Gerza-Ikhtiariya, and Korosko-Makhadma. This Basal group seems to have been deposited in a span of time which extended from 120,000 to 30,000 B.P.

The Dandara Formation (α Aggradational Silts and Fluviatile Sands)
The oldest of the aggradations, the Dandara, which marked the beginning of the Neonile, was accompanied by an arid episode which followed the Abbassia pluvial. Because the Dandara was followed by a long recession, it can be singled out by its patchy distribution, subjection to intense erosion and separation from succeeding aggradations by an exceptionally long interval of minor flow. As originally described (Said *et al.*, 1970), the Dandara was believed to underlie the Qena sands. However, as already explained, the Dandara silts are transgressive over the Qena sands and lie over their eroded surface with a depositional dip. The silts are similar, if not identical, to the modern silts except that they are hardened and include more calcareous materials. The type locality, as described by Said *et al.* (1970), lies at a point 3 km south of the temple of Hathor, Dandara, Upper Egypt, where a cut shows a unit of light brown, compact, and massive silty sand with occasional gravel bands of Eocene and Precambrian derivation and lenses of coarse sand. Plant casts are present. This unit is about 2 m thick (base unexposed). It is unconformably overlain by a 70-cm thick bed of rubble made up of gravels of different sizes and composition. The majority is of cobble size but occasional pebbles are also present. Several flakes and one core obtained from a depth of 1 m within the silt and under this cover indicate that the upper parts of the Dandara silts are of Mousterian (?) age.

A nearby section is described by Sandford (1934). The Dandara marl is described as purple, weathering red brick, veined with gypsum and with a 3-foot lenticle of rolled limestone rubble. It is given a thickness of 1.6 m and is described as lying unconformably over an irregular band of exceedingly hard rubble conglomerate which, together with the underlying

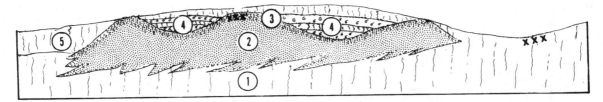

Figure 40. Hypothetical cross section of the older and younger Neonile deposits near Esna, Upper Egypt. After Wendorf *et al.* (1970).

5. Sahaba-Darau silts
4. Deir el-Fakhuri recessional deposits made up of two diatomites separated by a thin silt bed
3. Noncalcic soil
2. Ballana dunes
1. Masmas silts

"Pliocene" limestone, is quarried on a large scale. The underlying beds belong to the Issawia formation.

Between Dandara and el-Marashda (about 25 km downstream) brown marl layers appear below the surface gravel layers of the modern fanglomerates. Near Naga Hamadi also hard dark brown marl appears. To the west of Dandara, the unit reaches its maximum development (Figure 41). It is 15 m thick and is made up of a lower gray, loose, and fine calcareous sandy silt bed, and an upper bed of brown calcareous silt with a number of thin carbonate interbeds in the upper part of the section. A sample of the upper carbonate interbed is dated as more than 39,000 B.P. (Wendorf and Schild, 1976). The whole is covered unconformably by a veneer of gravel of local derivation including slightly rolled implements of Mousterian tradition. A thick red soil is preserved on the eroded surface of the formation below the gravel veneer. The limestone interbeds of the upper silt layers were deposited in a quasi lacustrine environment during the waning phases of the Dandara Nile.

In Nubia older silts which could be correlated with the Dandara seem to exist in many areas. They were previously described by de Heinzelin (1968) and lumped with other silts

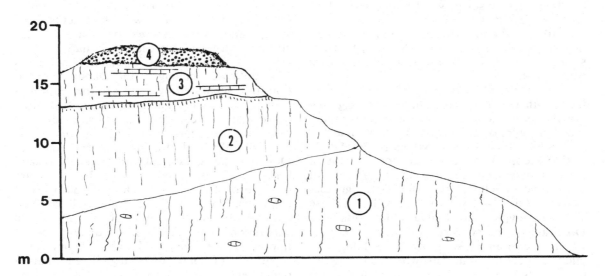

Figure 41. Section of hill along cultivation edge, 6 km southwest of Dandara, Upper Egypt.

4. Gravel in a matrix of coarse sand containing slightly rolled artifacts, ?Makhadma slope wash
3. Bands of freshwater limestone, the top layer gave a radiocarbon date of more than 40,000 years B.P.
2. Brown structured silt, hard with calcium carbonate concretions
1. Gray structured silt, hard and with thin sand laminae

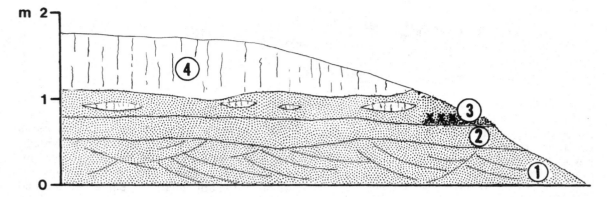

Figure 42. Site 1017 of the Combined Prehistoric Expedition, Khor Musa, Sudanese Nubia. After de Heinzelin (1968).
 4. Silt
 3. Living floor of the Khormusan industry
 2. Fluvial sand with silt nodules
 1. Cross-bedded fluvial sand

into his Dibeira-Jer formation. According to de Heinzelin, the Dibeira-Jer represented the oldest aggradational deposits of the Neonile characterized by a sandy facies rich in calcic impregnations (Dibeira) and characteristic accumulations of calcified roots and stems together with lenses of wind blown sand (Jer). At the inferred type locality of this formation (site 1017 of the Prehistoric Combined Expedition to Nubia, Figure 42) Marks (1968) describes a living floor of the Khormusan industrial complex.

The Dibeira-Jer had originally been correlated with the older Neonile silts (β aggradational silts in the present classification) recognized in Egypt by Said et al., 1970, Wendorf et al., 1970a,b,c. However, Wendorf and Schild (1976) reconsidered the sequence in light of their archeological findings in the β silts of Egypt and concluded that the older aggradational silts of Nubia were not coeval with those silts of Egypt. The Nubia older silts (the Dibeira-Jer) carry archeological materials which, according to Wendorf and Schild, have middle Paleolithic affinities, whereas the Egyptian β silts carry late Paleolithic materials. In addition, Wendorf and Schild questioned the validity of the radiocarbon dates which had been given in earlier publications (Wendorf, 1968) and which had led Marks (1968) to state that the Dibeira-Jer and its associated Khormusan industry covered the time span between 27,000 and 17,000 B.P. As further evidence, Wendorf and Schild cite Irwin et al.(1968) as giving older radiocarbon dates for the Dibeira-

Jer. These authors, therefore, would distinguish the Neonile silts of Egypt into a unit which they termed the Masmas-Ballana. The older silts of Nubia were considered as a complex unit.

A reexamination of the work of de Heinzelin shows that the oldest aggradational episodes of the Neonile in Nubia include silts which could be correlatable with the older silts of Egypt. In fact, the Nubia sections, unlike those of Egypt, have exposed bases of the older silts and outcrops of the underlying formations. In sections 440 and 1440 (Figure 43, a and b), cited by Wendorf and Schild as evidence for their argument, there are two aggrading silts separated by sand dunes and wadi sediments. The lower silt which underlies the dune is "very strongly hardened silt, almost as hard as concrete at the top." It carries calcic concretions and could well belong to the Dandara formation. It is separated by a marked unconformity from the overlying Ikhtiariya dunes. It is to be noted that silts carrying Khormusan industrial materials are recorded by Shiner et al. (1971) from the Wadi el-Milk area to the south of Dongola, the Sudan, under the name Goshabi formation. This formation in site N2-1 (Figure 44) is formed of fluviatile silt and sand overlaid by gravel. The fluviatile sands are well stratified and the silts are hard, compacted, and rich in concretions. The relief is inverted and gives the impression of being old. The unconformity separating it from the overlying Girra formation is marked.

Contrary to the Dandara section in Egypt, the section in Nubia carries fluviatile sands. In site 1017 of the Combined Prehistoric Expedition at Khor Musa, Sudanese Nubia, the section is made up of 6 m of fine-grained micaceous fluviatile sand which is cross-bedded and overlaid by brown structured and hardened silts. The absolute elevation of these sediments is 148–149 m; i.e., 26–27 m above the flood plain level prior to the erection of the Aswan High Dam.

In the Goshabi area (Figure 44) the lower fluviatile unit correlatable with the Dandara formation is made up of a 160-cm thick structured and hardened silt bed and an upper 120-cm thick bed of fluviatile micaceous sand with gravel interbeds.

It is interesting that whereas the vigorous Prenile cut its bed in the elevated Nubian massif leaving no trace of its sediments along its banks, the less vigorous Neonile deposited its sediments in Nubia in the form of coarse sediments becoming less coarse as it flowed northward into Egypt.

The Gerza-Ikhtiariya Formations

On the basis of a comparison between the thickness of the Dandara aggradational deposits and those of the later aggradational deposits of the Neonile, accumulating under a similar regimen and reasonably well dated, one can conclude that the Dandara episode lasted for no longer than 20,000 years. During this time, and for some time after the waning of the Dandara Nile, arid conditons prevailed over Egypt. This interval of aridity extended from the end of the Acheulian (Abbassia) to the Mousterian-Aterian pluvial (Korosko-Makhadma), an interval which continued for about 40,000 years.

During this arid episode gypseous soils de-

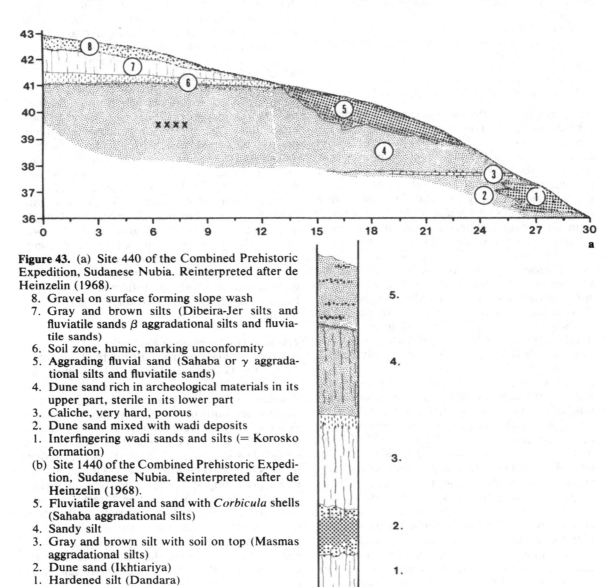

Figure 43. (a) Site 440 of the Combined Prehistoric Expedition, Sudanese Nubia. Reinterpreted after de Heinzelin (1968).
  8. Gravel on surface forming slope wash
  7. Gray and brown silts (Dibeira-Jer silts and fluviatile sands β aggradational silts and fluviatile sands)
  6. Soil zone, humic, marking unconformity
  5. Aggrading fluvial sand (Sahaba or γ aggradational silts and fluviatile sands)
  4. Dune sand rich in archeological materials in its upper part, sterile in its lower part
  3. Caliche, very hard, porous
  2. Dune sand mixed with wadi deposits
  1. Interfingering wadi sands and silts (= Korosko formation)
  (b) Site 1440 of the Combined Prehistoric Expedition, Sudanese Nubia. Reinterpreted after de Heinzelin (1968).
  5. Fluviatile gravel and sand with *Corbicula* shells (Sahaba aggradational silts)
  4. Sandy silt
  3. Gray and brown silt with soil on top (Masmas aggradational silts)
  2. Dune sand (Ikhtiariya)
  1. Hardened silt (Dandara)

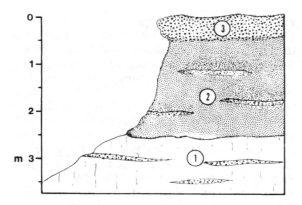

Figure 44. Profile at Goshabi, site N 2-1, Dongola, Wadi el-Milk, Sudan. Reinterpreted after Shiner *et al.* (1971).

Sahaba formation
3. Fluvial sand and gravel (mainly quartz and to a lesser extent jasper, etc.)

Goshabi formation
2. Fluvial sands
1. Structured silt with layers of gravel

veloped over the Prenile sediments and/or the Abbassia gravels. To the north and all along the western bank of the Nile, from north of Assiut to Cairo, the soil profile developing on top of the Abassia gravels is characterized by the presence of gypsum and salt pans of later genesis and probably of local derivation from the nearby rocks. A typical profile in this reach is in Darb Gerza where a gypsum pan developed to a depth and thickness that warrants its quarrying on a small scale. These gypsum incrustations occur just below the surface frequently at depths of 40–100 cm. Here the soil, termed Gerza, is mainly a gravel soil, though more often coarse sandy in part. The surface is gravelly desert pavement which overlies a sandy silt desert topsoil some 5 cm thick (color 7.5 YR 5/8 dry–5/6 moist), very rich in lime and platy in appearance. This topsoil is underlain by a 50-cm thick layer of gravel with very coarse sand in between, rich in salt crystals, and rich in lime having a color of 5 YR (5/6 dry–5/8 moist). Frequently a salt or anhydrite pan underlies this zone. Beadnell (1905) and more recently Abdallah and el-Kadi (1974) describe the Darb Gerza gypsum deposit and associate it with post-upper Pliocene sediments. Contrary to the views expressed here that this and similar deposits are postdepositional pans developing during an arid early late Pleistocene interval, these authors believe that this gypsum was deposited in lakes that were connected with the Mediterranean or with the remains of the Pliocene gulf of the Nile.

In addition to the gypseous soil, this arid interval was characterized by the accumulation of dune sand along the western banks of the river. In Nubia, where the base of the Neonile sections are exposed, dune sands of the Ikhtiariya formation underlie unconformably the Dibeira-Jer β aggradational silts and fluviatile sands. The sands of the dune are well sorted and fine to medium grained. The top part of the dune is truncated and topped by a humic soil. The dune rests unconformably over the Dandara silts in site 1440 of the Combined Prehistoric Expedition (Figure 43, b). The bulk of the dune has a number of caliche layers (Figure 43, a) indicating the highly arid conditions under which the dune was formed. The thickness of the dune varies from place to place but it reaches a maximum of 8 m in some places and averaging 4 m in most other places. The upper part of the Ikhtiariya dune carries archeological materials which belong to the middle Paleolithic.

Transgressing over the lower parts of the dune are wadi sediments which are described by de Heinzelin (1968) as "coarse, heterogenous (in composition) and stratified, poorly hardened except for calcic sheets." These wadi sediments could belong to the succeeding formation, the Korosko, which was formed during the Mousterian-Aterian pluvial. Judging from the descriptions of de Heinzelin (1968) it is possible that these wadi sediments interfinger rather than transgress the Ikhtiariya, and in this case the dune could be contemporaneous with the wadi sediment. It is indeed possible that the formation of the two sediments overlapped, for both carry faunas and archeological materials which belong to the Mousterian.

The Korosko-Makhadma Formations

A major pluvial occurred during the later intervals of the Ikhtiariya interval and continued for at least 60,000 years during the Mousterian-Aterian. During this pluvial man made a grand appearance both in the Nile Valley and in the deserts beyond. According to the most recent archeological work (Wendorf *et al.*, 1977) a relatively short interval of aridity occurred during this pluvial. The Mousterian-Aterian pluvial had a great impact on the landscape of Egypt. Intense erosion took place, the desert lands surrounding the valley were lowered and the modern landscape of the valley was developed (see cross sections of the valley, Figure 3, which show the disposition of the strata after their excavation in Mousterian-Aterian time). This pluvial also produced sheet wash deposits

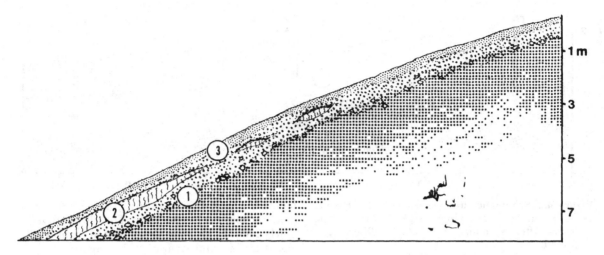

Figure 45. Section showing Makhadma slope wash developing over Qena sands at hill west of Makhadma village, Qena province, Upper Egypt.
  3. Upper slope, wash, unsorted fine- to medium-grained sand with small pebbles carrying late Paleolithic artifacts
  2. Silt (Sahaba formation)
  1. Lower slope wash (Makhadma formation) unsorted coarse-grained sand with pebbles, cobbles, and boulders cemented in a calcareous matrix; rolled middle Paleolithic artifacts are found

which rest unconformably over the slopes of the valley and many of the lowered surfaces of earlier formations. In this sheet wash rolled implements of middle Paleolithic tradition are known. The best-known of these occurrences is in Makhadma to the north of Qena where the slope wash overlying the Qena sands includes rolled implements of middle Paleolithic tradition (Wendorf and Schild, 1976). Other occurrences include the Abassia area from which Bovier-Lapierre separated Mousterian implements from the sheet wash overlying unconformably the Abbassia gravels. From the sheet wash overlying the Dandara formation, implements with middle Paleolithic affinities are separated by Wendorf and Schild (1976).

The Makhadma sheet wash deposits are thin and gravelly. They include pebbles of local derivation which could not have been transported for long distances. The pebbles are mostly in the 10–15 mm size range and are embedded in a matrix of yellow red soils in the 7.5 YR hue. The Makhadma deposits overlie unconformably the slopes and eroded benches of older deposits and underlie the sediments of the Neonile carrying fresh late Paleolithic artifacts (Figure 45).

In addition to the sheet wash deposits which develop over many slopes of the Nile Valley, there are the gravels and sands which accumulated in many areas on the east bank of the Nile below the Neonile deposits. Extensive terraces of alternating gravels and coarse sand occur at the base of the cliffs hemming the east bank of the Nile in Upper Egypt. These have been known for a long time and are amply described by Sandford and Arkell (1933) and Sandford (1934) from the Sebaiya area and Middle Egypt, respectively. Some of the sands carry an Ethiopian megafauna similar in many respects to that recorded from the Bir Sahara area, Western Desert, by Wendorf et al. (1977) such as *Equus asinus, Gazella rufifrons,* and *Hippopotamus amphibius.* In the Kom Ombo area Butzer and Hansen (1968) describe similar beds as the Basal sands and marls or the Korosko formation. In Upper Egypt, in Luxor and Sebaiya, Sandford and Arkell (1933, p. 84) note the presence of gravel and coarse sand beds in the 10-foot terrace; these include Mousterian artifacts. These authors remark that the implements were separated "from the basal part of the later deposits, which consist mainly of micaceous silts almost indistinguishable from that brought down by the Nile today." Wendorf and Schild (1976) note gravel and sand beds in all pits and trenches made on the east bank of the Nile, especially in the Dishna area, but they do not relate them to Butzer and Hansen's Korosko formation. Wendorf and Schild record, below the Neonile deposits, gravel and calcareous silt beds including in certain areas rare rolled artifacts of unknown age. They correctly call them wadi sediments, and even though they do not mention that these wadi sediments are separated by a surface of unconformity from the succeeding sediments, there is evidence that they are. In their attempt to cor-

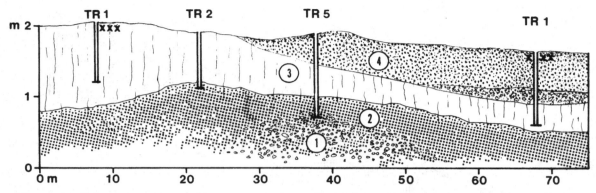

Figure 46. Schematic cross section at the Dishna plain, Qena province, Upper Egypt. Data of the trenches after Wendorf and Schild (1976), reinterpreted.

4. Upper wadi wash made up of cobbles and pebbles embedded in a sand matrix
3. Nile silt
2. Coarse sand wadi sediment grading underneath
1. Lower wadi wash made up of cobbles and pebbles of local derivation

relate the units encountered in the pits and trenches of the Dishna area, Wendorf and Schild assume the presence of two units of wadi sediments which they believe to precede and succeed the main Nile siltation in the area (Figure 46). The lower "wadi sediment" does not include artifacts and is separated by a marked unconformity from the overlying sediments. The upper "wadi sediment" includes in its upper layers artifacts which are closely related to the underlying main Nile silts of the Dishna area. It belongs to the Sahaba/Arkin episode of recession (= Dishna-Ineiba interval, see below). The lower wadi sediment is separated from the succeeding section by a surface of unconformity and is archeologically sterile. It belongs to the Korosko formation.

The lower wadi sediments of Wendorf and Schild (1976) described from the east bank of the river are obviously correlatable with Butzer and Hansen's (1968) Basal sands and marls (Korosko formation). These are properly distinguished by the latter authors as a unit "unlike any earlier or later deposits preserved in the Nile Valley of southern Egypt" (1968, p. 87). The unit is described as being made up of light-gray to light brownish-gray (2.5 YR 6/2) poorly sorted marly medium sand to sandy marl with about 15 to 40% $CaCo_3$. Two variant facies are recorded. The first, found below the marls, is made up of dispersed pebbles or local gravel concentrations, salt incrustations, and a high sand component. These marly sands with gravel indicate local wadi activity. The second sub-facies is a pale brown sandy gravel with abundant carbonates but no salts. The older gravelly marls seem to indicate depositon in local wadis and under a semiaquatic environment.

The Korosko, inspite of the above mentioned characteristics which clearly indicate significant contributions from the side wadis may have been influenced during short times by the running waters of the Neonile producing what Butzer and Hansen (1968, p. 96) described as "either a lacustrine or semiaquatic environment of deposition". The high carbonate content of the sediments "suggest precipitation in standing waters". The molluskan fauna which the Korosko carries (*Planorbis ehrenbergi, Bulinus truncatus, Lymnaea sp., and Corbicula fluminalis*), is taken to indicate subacqueous deposition in muddy waters.

Recent work along the Nile in Upper Egypt has now made possible the separation of a unit of gravels and marls below the Older Neonile deposits which is of areal extent. It corresponds to Butzer and Hansen's Korosko formation, a name that the present author proposes to retain even though he has not been able to locate the type locality which most likely has undergone under cultivation since the settlement of the Nubians in the Kom Ombo plain. The unit carries in places middle Paleolithic implements. Its age is given (Butzer and Hansen, 1968) as greater than 27,000 B.P. This is the Mousterian terrace *par excellence*. It was deposited by active side wadis during an episode of low Neonile. It is difficult at present to correlate this unit with any of the units recorded from Nubia, but if one accepts Wendorf and Schild's (1980) attribution of the Khormusan industry (known only in Nubia) to the late middle Paleolithic, it is possible that the Nile silts which include this industry (previously recorded as part of de Heinzelin's Dibeira-Jer formation) belong to the Korosko interval. However, the Khormusan silts of Nubia are alluvial in nature and, therefore, differ from the Korosko sediments. It is indeed possible that these Khormusan silts represented the earliest sediments of the Older Neonile deposits. The limited areal extent of these sediments and their disappearance under the waters of Lake Nubia makes the stratigraphic position of these silts difficult to place.

The Older Neonile Deposits

As previously stated, the older Neonile deposits form a sequence of nilotic deposits indistinguishable from those of the modern Nile. They are in the form of two aggrading silts, the β and γ which are separated by the recessional deposits of the Deir el-Fakhuri episode (β/γ deposits).

The disposition and stratigraphic order of the units of the older deposits differ from one place to another. In Nubia, which from a geological point of view extends to the gorge of Gebel Silsila north of the Kom Ombo plain, the valley is narrow and is hemmed for almost all its stretch by Nubia sandstone cliffs. Here, with the advent of the Neonile, downcutting took place continuously regardless of the volume of water received or the changes of local or ultimate base levels. The terraces left behind by the different aggradational episodes of the Neonile became progressively lower in elevation by time. The elevations of the terraces differ from one stretch to another because this youthful part of the Nile was obstructed by cataracts and waterfalls, some of which persist to this day. Among the cataracts that disappeared and can still be recognized from a geomorphic point of view are Batn el-Hagar (Sudanese Nubai), Madiq, Kalabsha, and Gebel Silsila (Egyptian Nubia) (Figure 47). The Silsila seems to have formed a waterfall until the advent of the Holocene; and as a result of its damming effect, the tectonic graben of Kom Ombo lying upstream was flooded and formed a swamp of considerable extent in the latter part of the late Pleistocene.

In Egypt, to the north of the Gebel Silsila gorge, the valley opens up with the result that the terraces of the different aggradational episodes assume different relationships from their corresponding equivalents in Nubia. The nature of the cross section of the Nile in each particular area determines the distribution, elevation, and relationships of the different terraces left behind by the different aggradational episodes.

The Masmas-Ballana Formation (β Aggradational Silts and Fluviatile Sands)

The β aggradational deposits of the Neonile complex overlie uncomformably the Korosko or the Ikhtiariya formations which carry Mousterian archeological materials, some of their

prototypes from the Negev Desert, Israel, have been recently dated by uranium series (^{234}U/^{230}Th) as 80,000 B.P. (Schwartz et al., 1979). It can, therefore, be stated that the basal parts of the β aggradational silts of the Neonile must have been deposited about 30,000 B.P. It must be recalled that radiocarbon dates for the middle layers of the β aggradational silts cluster somewhere between 21,000 and 17,000 B.P.

THE β aggradational silts of Egypt are amply described by Butzer and Hansen (1968), Said et al. (1970), and Wendorf and Schild (1976). Butzer and Hansen describe, from the Kom Ombo area in Upper Egypt, flood silts and channel beds which they term the Masmas formation. The thickness is estimated as exceeding 43 m. Terminal dates for the upper part of the silts are given as ranging from 17,000 to 18,000 years B.P. These silts are correlatable with the "lower silts" of Said et al. (1970) which outcrop in Upper Egypt, and also with de Heinzelin's (1968) Dibeira-Jer silts which overlie the Ikhtiariya dunes of Nubia.

The Masmas carries a rich molluscan fauna: Planorbis ehrenbergi, Bulinus truncatus, Valvata nilotica, Unio willcocksi, Corbicula fluminalis, and Cleopatra bulimoides. A full description of these and other fossil molluscs collected from these silts is given by Martin (1968).

Along the west bank of the Nile in Nubia and Egypt is a complex of dune sands which interfinger the upper layers of the Dibeira-Jer silts. These dunes are first recognized in the Ballana area in Egyptian Nubia and given the name Ballana formation by de Heinzelin (1968). A series of radiocarbon dates obtained from the top of the silts and the interfingering dunes in Egypt gives an age of between 18,000 to 16,880 B.P. which closely agrees with the dates of the type section of the Masmas formation in Kom Ombo. Wendorf and Said (1967) consider these dunes as representing episodes which accompanied aggradational episodes of the Nile; the stratigraphic evidence shows that the maximum of this aggradation occurred concurrently with the dune migration. For this reason Wendorf and Schild (1976) suggest naming the deposits of the dune and the interfingering Nile sediments the Masmas-Ballana formation. The Masmas-Ballana is rich in archeological remains of late Paleolithic tradition. Most of the archeological material comes from the top of the dunes as well as from their eroded surfaces (Phillips, 1973).

The Masmas and Ballana were contemporaneous deposits of the Neonile during the aggradational episode terminating at $\pm$ 17,000

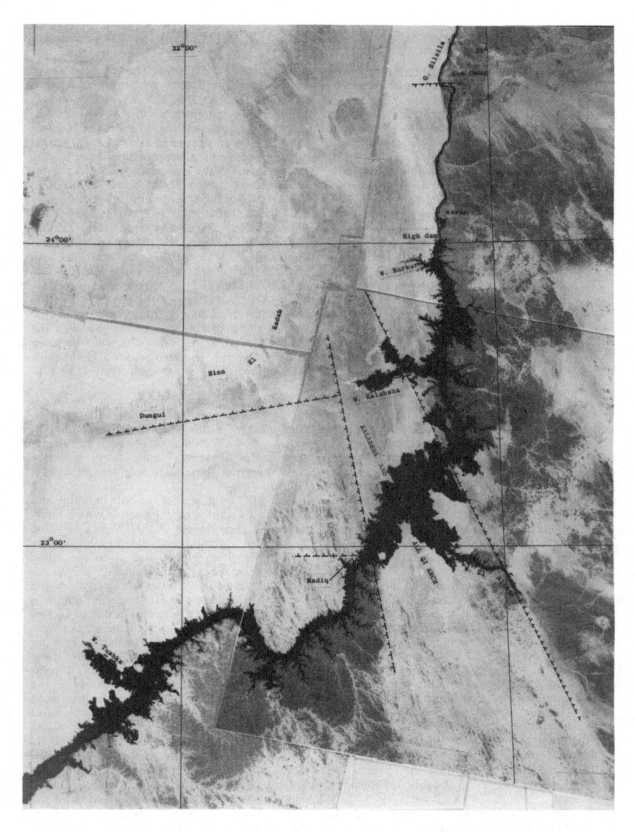

Figure 47. Combined space photographs of Nubia
showing ramifications of lakes Nasser (Egypt) and
Nubia (Sudan) filling ancient drainage channels.

B.P. They are correlated with the Wurm III/IV interval. The silts are typical flood-plain deposits. No wadi activity occurred during their deposition. The interfingering Ballana sand dunes indicate aridity and climatic conditions similar to those prevailing today. Indeed dunes interfingering Nile sediments along the western bank of the river in Egypt are observed to recur during the aggradational episodes of the Neonile as well as those of the Prenile. In the case of the Prenile, the dune sands interfinger the aggradational deposits of the Qena Prenile deposits along the west bank of the Minia-Beni Suef reach.

The Deir el-Fakhuri Formation (β/γ Recessional Deposits)

The Masmas-Ballana aggradation was followed by a recessional interval. The deposits of this interval are described from the Esna region by Wendorf *et al.* (1970c) under the term Deir el-Fakhuri formation. The deposits were formed in ponds that developed on the Ballana dune fields. They lie unconformably over the older dunes and are made up of two diatomite layers, each about 70 cm thick. In places they are separated by a silt layer about 20 cm thick. The Deir el-Fakhuri formation as originally conceived covers the entire series of recessional features as well as the intersiltation phase. The only other locality along the Nile Valley where the deposits of the Deir el-Fakhuri formation are recorded is in the Tushka area, Egyptian Nubia (Albritton, 1968).

The Deir el-Fakhuri overlies a noncalcic soil zone which developed over the stabilized Ballana dunes. The stabilization was aided by the formation of cultural debris and increased vegetation.

The level to which the Nile fell during this episode is unknown but probably around 24 m in Nubia (de Heinzelin, 1968), and probably less in Upper Egypt than the maximum attained during the Masmas-Ballana. Swales developed in the peripheral channels abandoned by the river and in the lows between the hummocks of the Ballana dunes. There is indication that the climate was slightly wetter than that prevalent during the aggradational episodes, for wadis draining the adjacent desert were active, eroding older deposits. However, the pollen and spores as well as the diatomite flora separated from the Deir el-Fakhuri formation indicate arid grassland environment (Wendorf and Schild, 1976). It is possible, therefore, that the Deir el-Fakhuri ponds got their water supply from the Nile itself by seepage, but one must assume that the summer termperatures were lower than those of today as the ponds seem to have stood for long periods of time.

Although a number of archeological assemblages are recorded from Deir el-Fakhuri formation and especially from the soil developing over the Ballana dune (Lubell, 1974; Wendorf and Schild, 1976), few radiocarbon dates are available from this unit. However, dates of preceding and succeeding events indicate that the formation was deposited within the span of 3000 years extending from 17,000 to 14,000 B.P.

The fact that a middle silt layer interfingers the Deir el-Fakhuri pond sediments leads Wendorf and Schild (1976) to believe that this succession may be looked upon as part of the succeeding aggradational episode. Consequently these authors lump the Deir el-Fakhuri formation with the γ aggradational silts. However, the sediments are clearly distinguishable in the field; and they were formed within a time span which, in other parts of the Nile, was accompanied by lower Nile levels and the formation of recessional features (Williams and Adamson, 1974).

The Sahaba-Darau Formation (γ Aggradational Silts and Fluviatile Sands)

Overlying the recessional features of the Deir el-Fakhuri formation are silts which attain an elevation of ca. 20 m above the modern floodplain in Nubia and ca. 8 m above this datum in Upper Egypt. These silts which are correlated with the Sahaba formation of Nubia (de Heinzelin, 1968), have a wide areal distribution in Egypt, and are the best preserved of the older Nile sediments.

The stratigraphic relationship between the older (Masmas-Ballana) silts and the younger γ aggradational deposits is significantly different in Upper Egypt and Nubia. The γ silts occur at a relatively lower elevation in Nubia and are inset against the older silts. The reverse relationship is seen in Upper Egypt. Here the γ silts occur at a slightly higher elevation above the older silts. This reversal may be due to the difference in base levels caused by the cataracts which once obstructed the path of the Nile in Nubia or to the dissimilar shapes of the cross-sections of the valley in Nubia and Upper Egypt, being narrow and frequently bound by cliffs in one instance and broad in the other. Regardless of what factor or factors are involved, it is evident that elevations must be used with great caution in identification and correlation of stratigraphic units along the Nile. Radiocarbon dates from within the Sahaba silts range from 12,060 B.P. (± 200 years) to 13,700 B.P. (± 300 years). For a discussion of these dates and their significance, the reader is referred to Wendorf *et al.* (1970b and 1979).

The Sahaba formation is correlatable with the Darau member of the Gebel Silsila formation of Kom Ombo (Butzer and Hansen, 1968). Wendorf and Schild (1976) propose to call this major, well-dated, and extensive aggradation, the Sahaba-Darau aggradation.

Some of the industries present in Nubia during the preceding Deir el-Fakuri episode survive into or through the episode of the Sahaba-Darau aggradation. These include the Qadan and the Sebilian, both of which seem to have been present in considerable numbers during this period. For a discussion of the archeology of the Sahaba in general and the Dishna area in particular the reader is referred to Hassan (1974).

As in the case of the Masmas-Ballana episode, the western bank deposits of the Sahaba interfinger with dune sands. This is particularly obvious in the el-Khefoug area where dune sands form the major component of the exposed Nile terrace. The dunes have long been stabilized. Considerable areas of these stabilized wind blown sands cover the fringes of the western banks of the Nile between Minia and Beni Suef over the lee of lower ground adjoining the valley bottom. On the higher land the sands often form elongated seif dunes. Their direction indicates the direction of the prevailing winds during their deposition which is similar to that prevailing today. Modern barchan dunes overlie the stabilized dunes unconformably and extend westward above the gravel and wadi wash of the neighboring areas. The stabilized dunes interfinger the Sahaba-Darau terraces with the result that the subsoils over large areas are sandy, with clay and silt layers penetrating deep in the profile. In fact, some profiles are rich in clay and can well be compared with the clay soils of the adjoining cultivated lands of the Nile Valley. Much of the el-Khefoug land is only periodically cultivated, and many of these sand soils have become loamy in the top layer through cultivation and irrigation. The mechanical composition of 20 samples from these dune sands indicates that most of them are made up of fractions less than 200μ (50.3%), $100-200\mu$ (30.8%), and $50-100\mu$ (17.2%). $CaCo_3$ percentages range from 4.6 to 8.4%. The type section of the el-Khefoug formation is north of Ihnasia el-Medina at the northern tip of this landscape.

The type locality of the Sahaba formation is in Nubia (locality 81 of the Prehistoric Combined Expedition, Figure 48). It is described by de Heinzelin (1968, p. 33, atlas Figs. 36, 37). Here the section has a different lithology from that known in Egypt. It is made up of a lower unit of strongly compacted and calcareous silt with sand and gravel intercalations. It contains rolled artifacts and fossils in poor condition. This is overlain unconformably by an upper unit of fluviatile sand with terminal gravel lenses. The sand and gravel contain a rich molluscan assemblage: *Cleopatra fluminalis*, many *Unio willcocksi* and *Corbicula vara*, fewer *Viviaprus unicolor* and *Valvata nilotica*, and some *Bulinus truncatus* and *Gyraulus costulatus*. The fluviatile gravel of locality 81 contains abundant artifacts and the pebbles are made up of brown quatzite, jasper, and chert. Vertebrate remains are abundant and are associated with the human occupation found in the silts. These are described by Churcher (1972, 1974) and Churcher and Smith (1972) from the Kom Ombo plain.

In Egypt the deposits of this aggradation are more uniform. With the exception of a few locally derived gravel beds that interfinger this aggradation, the Sahaba in Egypt represents a monotonous massive unit of silt with granulometric and mineral analyses which are similar to those of the modern flood plain silts of the Nile. Occasional gravel beds with rolled jasper pebbles are known in many localities especially in Luxor, Idfu, and north of Cairo in Sharkia province.

To sum up, about 30,000 years ago the river, with a regimen not very much different from that of today, cut its way into Egypt in a valley that had already been incised during the earlier pluvial. Prior to the advent of the older β Neonile many of the wadis of the east bank were choked by sediments of the Korosko formation, whereas many areas along the west bank had sand dunes and interfingering wadi sediments accumulating over the older α Neonile silts (Ikhtiariya). The Korosko wadi sediments and the Ikhtiariya dunes carry Mousterian implements.

The older β Neonile continued to build its bed (except for a short recessional interval, the Deir el-Fakhuri) to a level higher than the level of the present flood-plain by at least 12 m. This extended from the advent of the Neonile to the beginning of the Holocene, for a period of about 20,000 years of which less than 3000 years represented a recessional episode. During the

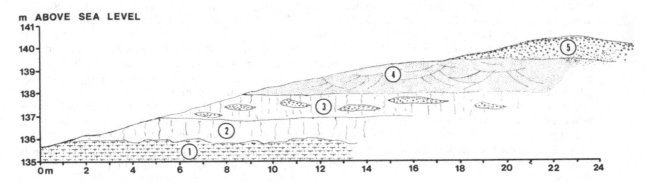

Figure 48. Site 81 of the Combined Prehistoric Expedition, Gebel Sahaba, Sudanese Nubia. After de Heinzelin (1968).

5. Gravel rich in jasper pebbles
4. Fluviatile micaceous sands, cross-bedded
3. Interbedded pebble beds
2. Structured silt
1. Bedrock, Nubia sandstone

aggradational episodes the river had floods which fluctuated in a manner not very much different from that prevailing in the recent past. During the recessional episodes the river became low or even wanting. During the first episode of normal (i.e., similar to present) floods silt and mud filled the valley gradually up to a level slightly higher than the present level of the flood plain. In many places the upper parts of these early deposits are recognizable in outcrop along the sides of the valley. During the succeeding aggradational episodes the river continued to build up its bed over the eroded surface of the preceding phase until it reached, in Sahaba-Darau time, the highest elevation known to have been reached by the Neonile. It was certainly in Sahaba-Darau time that the river had its maximum floods, and the column of sediments in Upper Egypt by the end of Sahaba-Darau time must have reached about 22 m in thickness, of which 12 m are now exposed.

The Younger Neonile Deposits

The post-Sahaba-Darau sediments of the Neonile are the deposits of the Holocene. They began after an episode of intense erosion (Dishna-Ineiba recession) which cut the valley from the heights it reached in Sahaba-Darau time to a depth which since then has been aggraded to the modern gradient of the valley. The cutting of the valley was probably in response to the eustatic sea level fluctuation which preceded the Holocene and which must have pushed the coast line much farther to the

north. The Holocene was an age of continuous aggradation. Except for the Dishna-Ineiba recessional episode which ushered the advent of the Holocene in Egypt, recessions represent minor events in an otherwise regular curve of accretion of the Nile silts over the valley and delta of the river. Indeed one can say that the regimen of the river (prior to its control) has been more or less constant for the past 9000 years and that the Nile in Egypt must have assumed at the inception of the Holocene its modern gradient.

The Dishna-Ineiba Formations (γ/δ Recessional Deposits)

The Dishna-Ineiba resessional episode is represented by playa and wadi deposits which interfinger the upper parts of the older Neonile deposits and the lower parts of the younger Neonile deposits. The deposits were formed during the major recession which separated the two aggradations.

Significant lower Nile levels are recorded in both Nubia and Upper Egypt, following the aggradation of the Sahaba-Darau fluviatile sediments. In Nubia few deposits may be referred to this recessional interval when the Nile fell 17 m from the maximum of the Sahaba aggradation. Only a few lenses of slope-wash debris at one locality appear to be of this age. This unit of slope-wash is named the Birbet formation by de Heinzelin (1968). There are no dates from this unit, but a sample from the base of the subsequent Arkin silts suggests that aggradation of that unit began around 11,200 B.P. (Fairbridge, 1962).

In upper Egypt this recessional interval is represented by playa deposits seen at several localities along the east bank of the river. In Dishna (Qena province) a well-preserved sequence of two playa deposits separated by a thin layer of silt is described by Said *et al.* (1970). These playa deposits accumulated behind the natural levees of the river in swales and abandoned channels of the Sahaba-Darau aggradation in the last phases of that event and during the subsequent recession.

The Dishna playa deposits suggest a similar environmental situation to that obtaining earlier when permanent ponds existed during the Deir el-Fakhuri interval. Again, it is not certain whether these ponds owed their existence to seepage from the adjacent Nile and/or to local rainfall. No archeological materials are found in association with the remnants of the Dishna playas.

Extensive playa deposits cover the eastern part of the Kom Ombo plain. Toward the east they overlie unconformably the Nubia sandstone bedrock, but toward the west the playa deposits interfinger the nilotic deposits of the Sahaba-Darau aggradation and overlie them. The thickness of the playa deposits is difficult to measure, but it exceeds 8 m. They are made up of a lower conglomerate bed composed of small pebbles of polygenetic origin embedded in a matrix of brown silts and fine-grained sands. These deposits are designated the Ineiba formation by Butzer and Hansen (1968) and described as "a widespread wadi accumulation the lower part of which is contemporary with Gebel Silsila Formation (= Sahaba-Darau) while its younger unit has no nilotic equivalent." The lower beds are designated the Malki member and these are coeval with the Dishna formation. The upper beds are designated the Sinqari member. The oxidation staining of the sediments and the presence of diffuse salts suggest deposition in an ephemerally standing body of water which was seasonally fed.

In the Atmur Nogra plain to the east of Kom Ombo, similar sediments are recorded. These cover an area of approximately 300 km². The Atmur Nogra plain itself must have formed a separate basin from that of the Kom Ombo plain; the Nubia sandstone wall which separated both basins was not breached except in post-Ineiba time. In spite of the fact that the Dishna-Ineiba sediments do not include archeological materials and are totally unfossiliferous, it is highly probable that the ponds continued to fill up during the final Paleolithic–Neolithic pluvial (9000 to 6000 B.P.) when the playa sediments interdigitated the younger Neonile deposits of Upper Egypt.

The Younger Neonile Deposits of the Valley and Delta

The younger Neonile deposits of the valley and delta of the Nile to the north of Nubia are buried. They have been accumulating since the Holocene forming a continuous column of sediments. The rate of sedimentation of the Nile mud over the lands of Egypt has attracted the attention of authors for a long time. Ball (1939) summarizes the earlier attempts and gives a detailed account of his calculation of the rate of accumulation of the Nile mud over the basin lands of Upper Egypt based upon a study of the fate of the suspended matter which enters Egypt at Wadi Halfa. Out of 110 million tons/year of suspended matter passing by this point, only 58 million tons pass by Cairo; the difference settles over the flood plain of the Nile between these two points. Ball estimates the rate of increase of thickness of Nile mud in the basin land of Upper Egypt prior to the erection of the Aswan Dam to be 10.3 cm/century. According to Ball, conditions obtaining in areas cultivated under basin irrigation approximate those of a natural river the flow of which is unhampered by irrigation projects and the flood waters of which are uncontrolled by man.

No similar studies to those of Upper Egypt are available, and probably none will ever be made, for the rate of sedimentation of the Nile silts in the delta. The lands of the delta were converted to perennial irrigation at the beginning of the last century prior to the establishment of a scientific body to record the amounts of suspended matter carried past the canals. The sedimentation rates for the perennially irrigated lands are much lower than those for basin irrigation, as demonstrated by Ball (1939) who finds that the amount of Nile mud deposited in the perennially irrigated lands of Upper Egypt is about 30% less than that in basin lands in the same district. The thorough control of the water going into the delta distributaries since 1833 (date of erection of the Barrages, north of Cairo, upstream of the bifurcation of the distributaries of the delta) has doubtlessly reduced the deposition of silt in the delta. Ball (1939) finds that the controlled perennial irrigation schemes in the delta have reduced to a minimum the accumulation of Nile muds there. He estimates the rate of deposition in a fully controlled river in the 1920s to be only 0.6 cm/century.

In a natural delta environment the surface of the water in the distributaries is lower than that in the river resulting in a swampy environment, overflow of the water from the channels, and a greater rate of sedimentation. The lower surface of water in the distributaries of a natural delta can be inferred from the water levels today in Cairo and the Barrages. Here the fall of the water surface between these two points is considerable; the difference between the mean water levels of the river at these two points is 1.2 m. Perhaps the closest figure for the rate of sedimentation in a natural delta environment could be obtained by comparing the thickness of the column of Neonile sediments in the delta with that in the valley, assuming that both are contemporaneous. Both form the agricultural layer of Egypt and both overlie unconformably the Prenile graded sand–gravel bed. The Neonile sediments in the delta are slightly thicker (Figure 49). By comparison with the rate of sedimentation for the valley, which was calculated to be 10.3 cm/century, the rate of sedimentation for the delta would be 12.5 cm/century or twenty times that prevailing today in the perennially irrigated delta lands.

As previously stated, the amount of suspended matter that passes Wadi Halfa is 110 million tons of which only 25 million tons flow to the Mediterranean to build up the delta cone. If this figure is projected back in time to the beginning of the Neonile system, then the share of the Neonile sediments in the total volume of the sediments of the delta cone would be negligible indeed. The Neonile sediments could have hardly formed except a thin layer on top of the delta cone.

The fact that the valley and delta were aggrading throughout the Holocene makes the finding of outcropping deposits of this episode

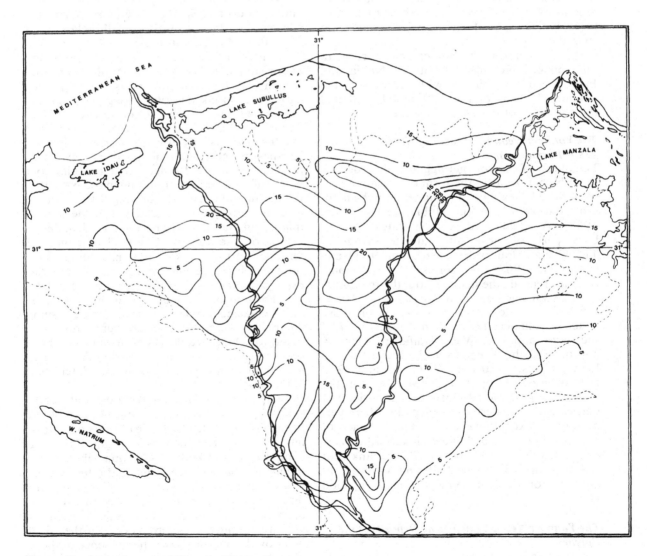

Figure 49. Isopach map of Neonile (Q$_3$) sediments.

within the Nile Valley and delta most unlikely. The places where the Holocene Neonile sediments can best be examined in outcrop are in Nubia and the Faiyum province. There is evidence that degradation was active during the Holocene over most of the stretch of Sudanese and Egyptian Nubia where numerous cataracts interrupt the river and many more must have been present in the very recent past. The Arkin sediments which represent the aggradational deposits of the Holocene in Nubia are found at a level of 13 m above the modern flood-plain in Nubia (prior to the erection of the Aswan High Dam) becoming gradually lower with minor stands by time. The flooding of the Faiyum depression by the Neonile at certain intervals left behind lacustrine sediments and beach features which can be examined in outcrop. In addition, the delta coastal lands show many features which were fashioned by recent and subrecent forces.

It may, therefore, be convenient to deal with the younger Neonile deposits of Nubia and Faiyum and terminate with notes on the younger deposits of the continental shelf and the evolution of the delta coast line.

The Younger Neonile Deposits of Nubia (The Arkin Formation)
The oldest of the younger aggradational deposits of Nubia occur in the form of embankments of silt which lie about 13 m higher than the modern flood-plain in Arkin, Nubia (de Heinzelin, 1968). Available radiocarbon dates for the basal parts of these silts indicate an age of 11,200 B.P.. (Fairbridge, 1962) and 9400 B.P. (Chatters, 1968). A series of recessional beaches records the decline from the maximum of the Arkin aggradation in Nubia. A number of radiocarbon dates from sites located on these beaches from near Dibeira, on the west bank of the Nile opposite Wadi Halfa, provides a chronicle of this event from the highest to the lowest levels: 7750 B.P., 5650 B.P., and 5270 B.P. In Egypt proper, the sediments of the younger aggradational episodes are buried underneath the modern alluvial plain. However, during one exceptionally high flood a bench of modern silt is recorded. Near el-Kab, Idfu, Vermeersch (1970) describes a silt embankment which is dated between 7980 and 8400 B.P. at an elevation of 4 m above the modern flood plain.

The Younger Neonile Lacustrine Deposits of the Faiyum
The Faiyum is a circular depression in the limestone plateau of the Western Desert to which,

in the course of time, the Nile waters obtained access through the Hawara channel. It formed an escape for the river waters especially during flood times in certain intervals of the Holocene. Nile sediments fringe the depression on its sides and provide evidence of its past history.

The Nile deposits of the Faiyum depression have attracted the attention of authors for a long time. Opinions differ as to the time when the Nile waters obtained access to this depression and as to the extent and level of the lake which once occupied the depression and which is today represented by Lake Qarun. Among the workers who contributed to our understanding of the fluctuations of the lake mention is made of Beadnell (1905), Caton-Thompson and Gardner (1934), Little (1935), Huzayyin (1941), and more recently Said et al. (1972a,b) and Wendorf and Schild (1976). Caton-Thompson and Gardner, as well as Ball, believe that the Nile had access to the Faiyum as early as Paleolithic time and that a lake formed in the depression and filled it to a level of +40 m (the present-day level of the lake is −43 m). This early lake shrank marking a long series of different lake levels from +40 m downward to −50 m when the Nile connection became severed and the lake possibly dried up. Early in Neolithic times the Nile again obtained free access to the depression and formed a lake reaching to a height of about +18 m. Later on in the Neolithic period, the lake level fell by stages to about −2 m owing partly to the reduction of the annual influx from the river and partly to gradual desiccation of the region. Caton-Thompson and Gardner further believe that pastoral-agricultural Neolithic people immigrated to the depression about 6000–5000 B.C. when the lake stood at a level of about +18 m; but that later on, increasing dryness of the climate having compelled them to give up agriculture and resort to fishing as a means of sustenance, these people followed the lake as it shrank to −2 m level and eventually dried up not later than about 4500 B.C.

Said et al. (1972a) and Wendorf and Schild (1976) believe that the older 44- and 40-m "beaches" recorded by Little (1935) could be of fluvial rather than lacustrine origin. They are composed of a mixture of sand and pebbles containing Corbicula, rare Unio, and a few water-worn artifacts. The shells of Corbicula are disarticulated but intact and show no evidence of rolling, while the gravels are nilotic in aspect and are rounded to subrounded. Further field work by the author shows the presence of these old terraces at a more or less constant elevation along the northeastern and southwestern rim of the depression. The terraces are associated with the depression which must have been in existence as early as Prenile time. In both the northeastern and southwestern stretches the surface of the deposit developed in many places a gravelly gypseous soil such as that described earlier from Gerza. A long period separated these early deposits from the succeeding lacustrine sediments, which belong to the Holocene, and it must be assumed that the Nile had no access to the depression for a period of about 130,000 years.

The Holocene lacustrine deposits of the Faiyum occur as embankments (Arabic *Gisr*) skirting the depression. The most continuous of these are those which lie at 18 and 23 m above sea level.

Said *et al.* (1972a, b) show that the first connection between the Nile and the Faiyum in this sequence occurred as late as 9000 years ago when thick diatomites were deposited. This was followed by a period when the lake shrank and the connection was severed to be established again only around 8150–8120 B.P. when the lake stood at an elevation of + 17 m. During the following 600 years (around 7550 B.P.) its level fell to + 12 m and then rose during the following 340 years (7190 B.P.) to a level of + 19 m or even as high as + 24 m. There is indication that the connection with the Nile was then severed and that the lake shrank to very low levels. The connection was again established after 1250 years (5910 B.P. ± 115 years) when Faiyum A people settled around a lake which had an elevation of + 15 to + 18 m. The lake continued to rise with minor fluctuations until it reached its maximum in Old Kingdom times (4800 B.P.). The shores were elaborately revetted, and a temple (Qasr el-Sagha) was built overlooking the lake. The later history of the lake is treated by Shafei (1940) who concludes that the lake's level was artificially reduced by the engineering work of Amenemhat I (1980 B.C.) who kept it at a level of + 18 m. Ptolemy I lowered and controlled the level of the lake by building a barrage at its entrance in el-Lahun in order to reclaim the silt fan left behind by the lake.

Figure 50 is a geological map showing the distribution of the Quaternary deposits in the Faiyum province.

The oldest lacustrine sediments are in the form of two diatomites each about 3 m in thickness. The mineralogy and faunal content of these beds are worked out by Kholeif (1973). Samples of the upper bed have a $CaCo_3$ content of 60% while those of the lower bed have a content ranging from 20 to 34%. The sediment is highly porous and is formed of mud or ooze carrying diatoms belonging to the following species: *Cocconers placentula, Cymbella ventricosa, Rhopalodia gibba, R. rhopala, Epithemia zerex, E. zebra, Fragilaria brevistriata, Stephanodiscus astraea,* and *Synedraulua* sp. The diatoms are similar to those living today in Lake Afar, Ethiopia. Many species are cosmopolitan, but *Rhopalodia rhopala* is an African species.

The accompanying figures depict the stratigraphy of these lacustrine sediments in the Faiyum area (Figure 51).

Wendorf and Schild (1976) believe that the Nile obtained access to the Faiyum in periods of exceptional high floods. They specifically mention the time of el-Kab higher terraces (Vermeersch, 1970) and that of Catfish cave in Nubia (Wendt, 1966) as contemporaneous with the deposits of the Neolithic and later sediments in the Faiyum. In the case of Catfish cave, silts lying 18 m above the flood-plain level are associated with a radiocarbon date of about 7100 years B.P. The first access of the river to the depression coincides probably with the Arkin aggradation.

The earlier history of the Faiyum depression is not known in any detail. The depression forms a basin with no external outlet. Borings in the Lahun gap by the Geological Survey of Egypt hit bedrock at an elevation of −17 m (Little, 1935) while the lowest point in the depression is close to −43 m. Most authors attribute the origin of the depression, therefore, to wind action; but recently Said (1979) advocates a tectonic origin for this and other depressions in northern Egypt and considers them as crustal sags in the elevated sea floor of the Mediterranean. The first break through into the depression occurred most probably in the latter part of the Prenile time, but evidence here is incomplete. However, it is certain that the Neonile did not have access to the depression except during the Holocene, even though the river during the late Pleistocene was high enough to flow into the depression. One must assume, therefore, that the Lahun gap was silted up during this time. During the Dishna-Ineiba recessional episode the gap was cleared of the material which silted it up and assumed a depth which allowed the waters of the δ Neonile to flow into the Faiyum depression. But the access was of short duration as the gap silted up again and was not bridged over except during exceptionally high floods. It was only during the Pharaonic period that the depression received an annual flow; and it is not clear whether this was due to the artificial periodic cleaning of the

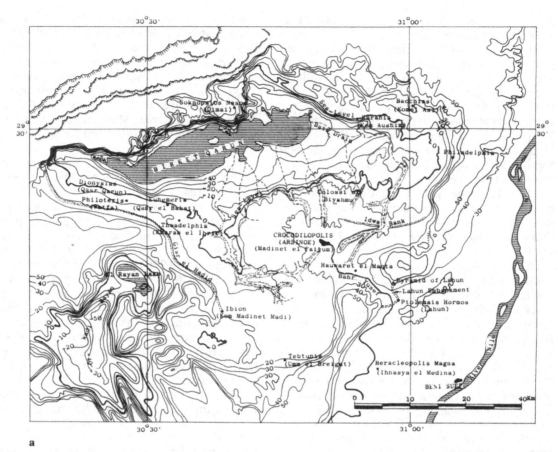

a

Figure 50. (a) Map of the Faiyum showing topography and sites of old towns. (b) Map of the Faiyum showing Quaternary geology. B, basalt; Te, Eocene, Tem, middle Eocene (Ravine beds); Teu, late Eocene, To, Oligocene; Tm1, early Miocene; Tpl, Pliocene; Q_1, Protonile Idfu formation; Q_2, Prenile lacustrine beaches; Q_3, Neonile lacustrine sediments; Q_d, sand dunes; Q_w, Recent wadi alluvium.

gap from the silt or to the fact that the Neonile had a series of exceptionally high floods during Old Kingdon times.

The Younger Neonile Deposits of the Continental Shelf and the Evolution of the Delta Coastline
The morphology, hydrography, and sediments of the continental shelf off the Nile delta are dealt with by a number of authors (Aleem, 1972; el-Wakeel and el-Sayed, 1978; Summerhayes *et al.* 1978; Misdorp and Sestini, 1976a,b). Misdorp and Sestini describe the continental shelf off the Nile delta as made up of a series of terraces separated by low slopes that are cut by drowned channels and by one major submarine canyon (the Rosetta canyon). The most extensive of these terraces are the

so-called upper and lower terraces breached by the Rosetta and Damietta cones. The slope breaks of the shelf occur at 14–18 m, 23–25 m, 41–45 m, 72–81 m, and 119–121 m. Whether these breaks are the result of erosion or deposition in the interval of standstill of the continuously rising sea level during the Holocene is not known. It is interesting that the edge of the upper terrace corresponds to the standstill at 19–20 m (5800–6400 B.P.), while the surface of the lower terrace to the standstill at 43–45 m (8000–10,000 B.P.). Slope breaks and terraces at depths similar to those on the Nile shelf occur in other parts of the Eastern Mediterranean coasts (Emery and Bentor, 1960) and elsewhere (Milliman and Emery, 1968). The conspicuous break at 121 m is at the same depth as the average for all continental shelves as calculated by Shepard (1964) and may represent the lowest sea level during the Wurm regression. It is to be noted that the shelf breaks are relict features which seem to have passed through a complex history. Neev (1968) finds that some of the terraces of the coast of Israel are rimmed by ridges of cemented "kurkar" (eolian sands) and the rest of the surface is

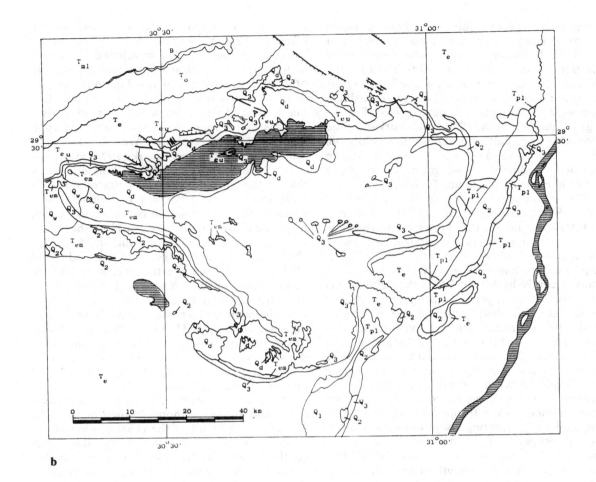

b

Figure 51. Section north of Qasr el-Sagha temple, type section of youngest Faiyum Moeris Lake sediments. After Said *et al.* (1972a).

9. Sand, fine-grained, compacted with a 7-cm thick carbonaceous layer at base
8. Sand, medium-grained, friable, light brown, caliche nodules
7. Sand, fine-grained, mottled
6. Sand, medium- to fine-grained, contains abundant snails: *Bulinus truncatus, Lymnaea lagotis,* etc.
5. Sand, medium-grained, reddish brown
4. Sand, fine-grained, mottled
3. Silt, bedded with concentration of gastropod shells *Bulinus truncatus, Bithynia conollyi, Planorbis planorbis,* and *Lymnaea* sp.
2. Sand, fine-grained, laminated with root casts
1. Diatomite, gray, calcareous, evenly bedded with gastropod shell fragments along bedding plains

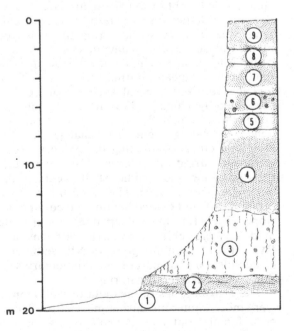

covered by recent sediments. This situation seems to exist also in the northern reaches of the delta. Drillholes in the northern reaches of the Beheira province show oolitic sand layers, "kurkar," at depths varying from 7.5 m (Kafr el-Dawar station) to 106–124 m (el-Taftish estate west of Rosetta). A description of these logs is found in Attia (1954). The kurkar oolitic limestone in el-Taftish boreholes lies below the Prenile graded sand–gravel unit and above the Paleonile sediments, and probably marks a shore-line during the lower Pleistocene.

The sediments of the continental shelf off the delta (Summerhayes et al., 1978) are mainly terrigenous sands on the shore face, terrigenous muds on the middle shelf, and algal carbonates on the outer shelf. Contrary to the ideas expressed by Sandford and Arkell (1939) and more recently by Shata (1971) the delta seems to have been larger by the advent of the Holocene. The distribution of the sediments leads one to conclude that the coast of the delta was retreating throughout the Holocene and that the coastline was about 40 to 50 km away at more or less the 100-m bathymetric contour line. About 10,000 B.P. the drainage system was drowned by the transgression. From then on fine sediments were trapped in depressions in the landward retreating coastal system, deposition of terrigenous muds more or less ceased on the fan, and the waters of the outer shelf became clear for the growth of coralline algae. Over the past 5000 years the sea level has been stable and during most of this time, prior to human interference, several distributaries discharged sediment through an arcuate, wave-dominated delta front; westerly longshore currents similar to those occurring today forced plumes of suspended sediment east along the coast. The reconstruction of the paleoclimate of the Mediterranean during the late Pleistocene–Holocene interval, such as that recently attempted by Thiede (1978), must take into account the fact that there is no geological evidence indicating significant changes in the regime of currents affecting the Eastern Mediterranean during this interval. Muds were distributed along the middle shelf, keeping the outer shelf free of fines. The modern Nile cone sediments do not extend beyond the continental slope. The Herodotus deep does not include modern Nile clastics (Venkatarathnam and Ryan, 1971), even though older Nile sediments (Paleonile) build up part of the sedimentary section of the Mediterranean ridge.

The distributaries of the delta of the Neonile were more numerous during most of the Holocene fanning out as far eastward as the old Pelusiac branch and as far westward as the Canopic branch (Figure 52). Seven major branches of the delta are mentioned in various historical documents and in ancient maps. Five of them degenerated and silted up in the course of history, whereas two, the present-day Damietta and Rosetta branches, remain active. The history of these branches is summed up in Tousson (1923). Sneh and Weissbrod (1973), Sneh et al. (1975), and Bietak (1974) study the defunct Pelusiac branch of the Nile and show that in the Sinai stretch the delta of the Nile built up its front in the Bay of Pelusium as a result of the accumulation of sediments which had been moved by the eastward longshore currents. Sneh et al. (1975) show that in 25 A.D. the shore-line off the Pelusiac branch stood about 10 km inland from the present shore-line. In this respect the easternmost part of the delta is in sharp contrast to other parts of the delta coastline which are known to have been retreating since at least the Holocene, first in response to the eustatic rise of sea level and then, when the sea level stabilized around 5000 years B.P., by coastal erosion (Said, 1958; Misdorp and Sestini, 1976a,b; Sestini, 1976).

Along the westernmost part of the delta coast, the ancient Canopic branch of the Nile silted up as a result of the reexcavation of the Bolbitic canal which today forms the upper reaches of the Rosetta branch. The reexcavated canal has less meanders and a greater gradient than other branches of the Nile, thus usurping a large part of the water passing through the bifurcation of the delta branches to the north of Cairo. This has resulted in the gradual silting up of the other branches of the delta. The Rosetta branch receives even today more than 70% of the water of the Nile as it bifurcates into the delta fan. Indeed, had it not been for the continuous efforts of the Egyptian irrigation engineers, the Damietta would have silted up long ago. The Bolbitic was excavated during the fifth or sixth century B.C.

The exceptional quantity of water which goes into the Rosetta branch has converted the delta from a Niger-type delta (Wright and Coleman, 1973) to a Mississippi-type delta where promontories are evident around its two surviving tributaries and especially around the Rosetta branch. Figure 53 is a schematic diagram showing the evolution of the deltaic coast line from an arcuate smooth line to a "bird-foot" line.

The Modern Nile

It may not be out of place to terminate these notes on the Neonile by giving a brief description of the most salient features of the topography and hydrology of the modern Nile. The literature on the modern Nile is extensive and the following notes touch only those aspects of the river that may be of interest to a better understanding of the past behavior of the river, its evolution, and its utilization by early man.

The course of the Nile from the mouth of the Atbara to the Mediterranean Sea may be roughly divided into two parts: the first stretching as far as Aswan where the river flows in a rocky channel with alternations of rapids and reaches of more gentle slope and where, on the whole, it is eroding its bed; and the second from Aswan to the sea where it traverses its own floodplains. In neither portion does it receive any appreciable addition to its volume, but owing to percolation and cultivation it is continuously losing water by amounts of which little is known. In connection with this it should, however, be noted that seepage through the bed into the surrounding areas is by no means in the nature of a dead loss as great portions of the water which is thus drained from the river at its higher stages (when its level is above that of the surrounding water table) percolates back again during the lower months when conditions are reversed. Thus, the discharge reading at Aswan prior to the erection of the artificial means of flood control during the low months is larger than that at Wadi Halfa. This is an extremely important point with regard to the regimen of the river in past geological episodes.

Figure 54 gives the slope of the Nile from Khartoum to the Mediterranean. It is to be noted that the Main Nile from Khartoum to Aswan falls 295 m in 1810 km; the so-called 6 cataracts occupy 565 km with a slope of 1:3000 and a total drop of 192 m, and the ordinary channel occupies 1245 km and has a slope of 1:12,000 and a total drop of 103 m. From Aswan to Cairo, in a length of 970 km, the Nile falls 76 m with a mean slope of 1:13,000.

The velocities of the river in flood and low supply are greatest in the cataract region (averaging 2.3 m/sec and 1.2 m/sec, respectively); they are lower in other reaches of the Main Nile (e.g., 1.7 m/sec and 85 m/sec between Aswan and the Mediterranean). The time the river takes to traverse the different reaches differs from place to place. From Khartoum to Aswan the Nile takes 11 days in flood and 22 days in low supply; from Aswan to Cairo it takes 6 days in flood and 12 days in low supply.

From Aswan to the Barrages (North of Cairo) the length of the river is 973 km in flood. The slope in summer is 1:13,000 and in flood 1:12,000. The mean fall of the valley is 1:10,800. The slopes vary in the different mean reaches, the least being 1:14,800 in Qena province and the greatest being 1:11,400 in Beni Suef. In a high flood with a rise of 9 m in Aswan, the rise in Qena is 9.5 m and only 8.2 m in Beni Suef. Neglecting spill channels, we may state that in a high flood the mean area of the Nile is 7,500 m² and the mean width 900 m. In the province of Qena the area is 7000 m² and the width 800 m, while in Beni Suef the mean area is 8000 m² and the mean width 1000 m. Speaking generally, it may be stated that where the Nile Valley is narrow the slope of the river is small, its depth great and width contracted; while where the valley is broad the slope is great, the depth small, and the width enlarged. It may be said that the Nile in summer has a natural section whose width in flood is 100 times its depth, while its mean velocity is 1.50 m/sec.

Past Cairo the river fans up into the delta. Figures 55 and 56 (after Willcocks, 1904) give the data of the bed of the Rosetta and Damietta branches, their ground levels, flood levels of 1892, tops of their banks, and slopes of the water surface. These figures show that the Rosetta branch in its middle reaches is from 1.5 to 2 m above the level of the country in a high flood, and that the Damietta branch is from 2.5 to 3 m. The figures also show that the slope of the first part of the Damietta branch is considerably less than that of the Rosetta branch; this results in the gradual silting up of the former. This may also help explain the greater thickness of the agricultural clay layers along the Damietta branch (Figure 49) and the lower equivalent blackbody temperature (T_{bb}) along this branch of the delta as recorded by the Nimbus Meteorological Satellite (Figure 57, after Sabatini et al., 1971), for here water retention would be at a minimum.

In spite of the great strides that were made in the past hundred years with regard to the study of the Nile, its water balance, especially after the building of the great Aswan Dam, is not yet fully understood. Even more difficult to unravel is the hydrology of the natural river unhampered by man's structures and interference. Since the dawn of civilization in Egypt about 5000 years ago, the problems of water storage and flood control have occupied the attention of every enlightened ruler of Egypt. During this long history the harnessing of the river has ranged from the laying down of gigantic blocks on both banks of the river, to the strengthening of the natural levees of the Nile, to the erection of the great structures of the water schemes of the nineteenth and twentieth centuries. These, as well as the continuous interference by man

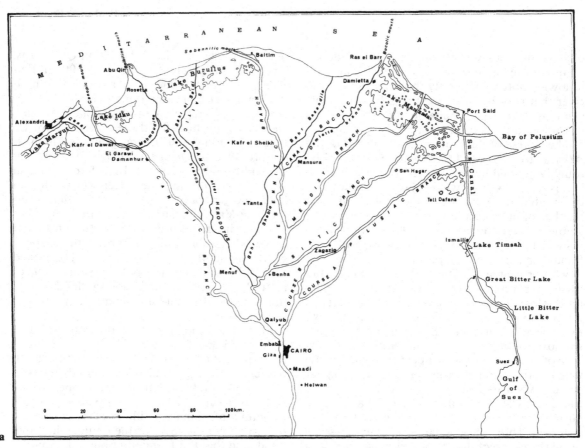

a

Figure 52. Map of the Nile delta showing ancient defunct distributaries. (a) After Herodotus (484–425 B.C.); (b) after Strabo (63 B.C.); (c) after Serapion (350 A.D.); (d) after El-Idrisi (1099–1154 A.D.).

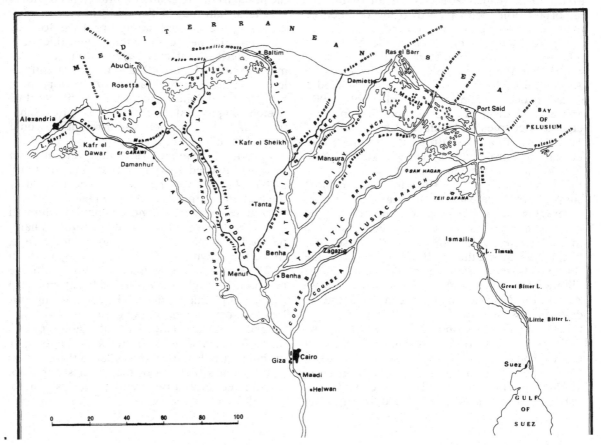

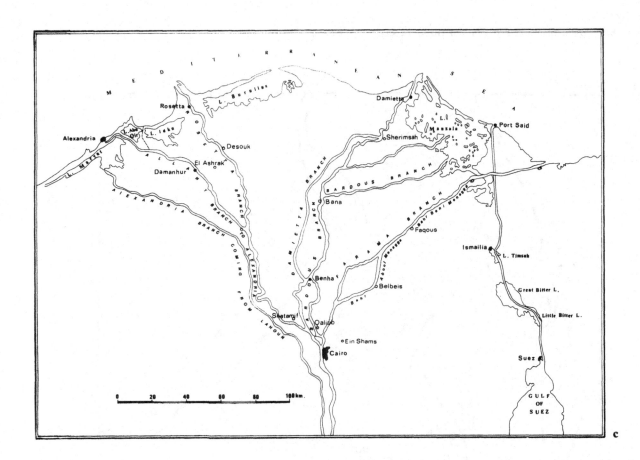

c

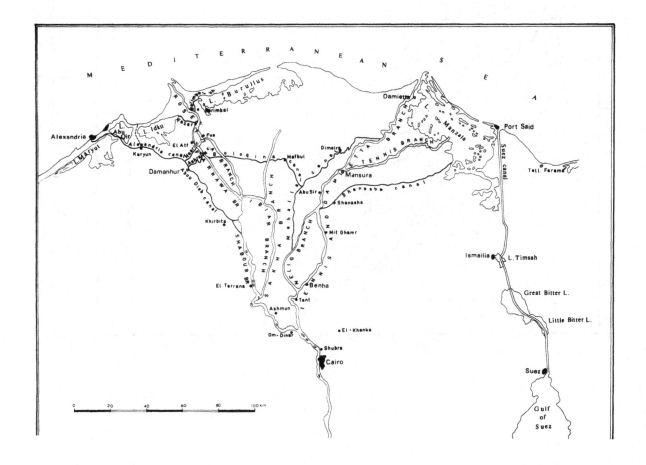

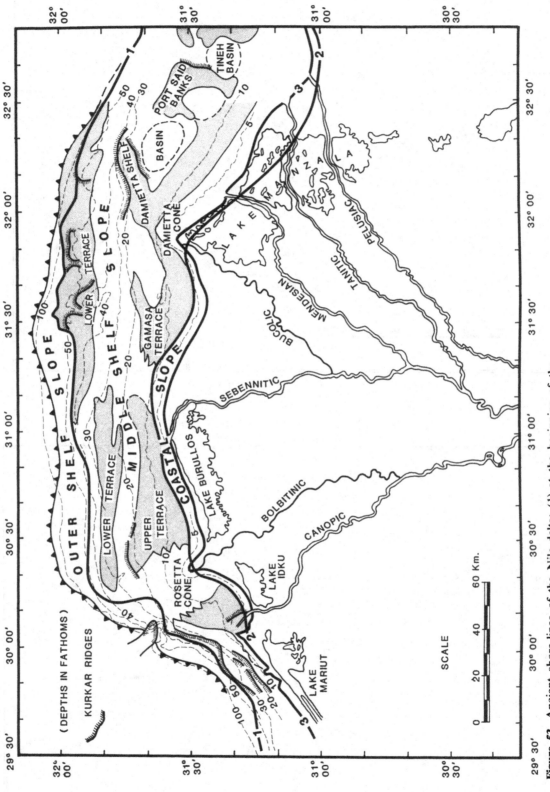

Figure 53. Ancient shore-lines of the Nile delta: (1) at the beginning of the Holocene, (2) in historic times, and (3) in modern times. Morphological units of the continental shelf of the Nile delta are after Sestini and Manohar (1976).

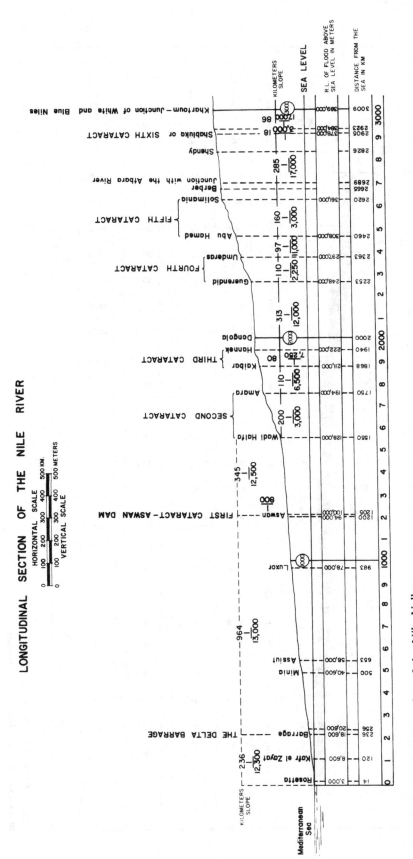

Figure 54. Longitudinal section of the Nile Valley from Khartoum to the Mediterranean. After Willcocks (1904).

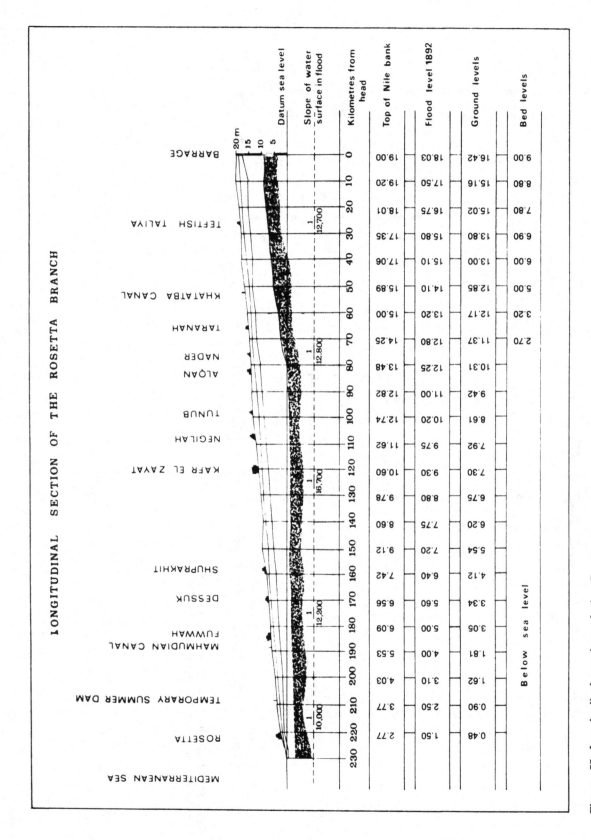

Figure 55. Longitudinal section of the Rosetta branch of the Nile delta. After Willcocks (1904).

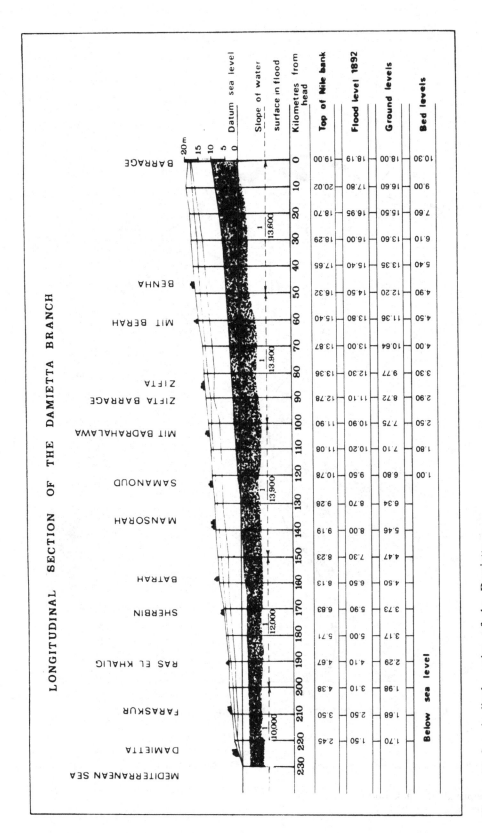

Figure 56. Longitudinal section of the Damietta branch of the Nile delta. After Willcocks (1904).

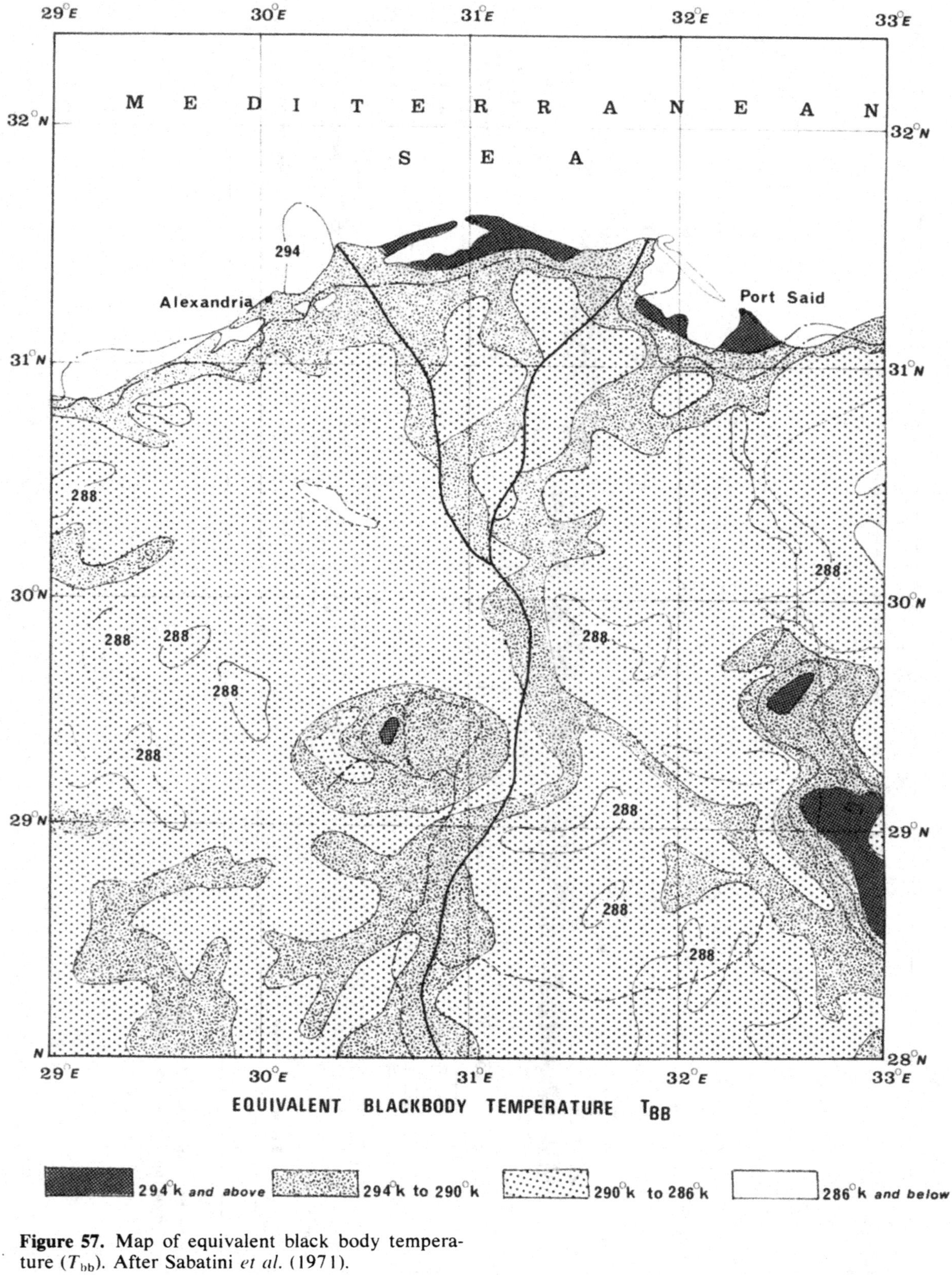

EQUIVALENT BLACKBODY TEMPERATURE T$_{BB}$

294°k and above   294°k to 290°k   290°k to 286°k   286°k and below

Figure 57. Map of equivalent black body temperature (T_{bb}). After Sabatini *et al.* (1971).

through wise or unwise measures, make it difficult to understand the natural regimen of the river. The following notes attempt to summarize the salient features of the nearest system to the natural regimen of the river, i.e., before the filling of the Aswan reservoir in 1902. For full details of the hydrology of the river and the effects of the building of the reservoirs on its regimen the reader is referred to the classical works of Lyons (1906), Garstin (1904), Willcocks (1889, 1904), Willcocks and Craig (1913), and the compendium on the Nile written by Hurst *et al.* of which volume 19 appeared in 1966. The hydrology and biology of the river is also discussed by Rzoska (1976).

The Egyptian Nile derives its waters almost exclusively from the rain which falls over two elevated areas: the Equatorial plateau of Central Africa and the Ethiopian plateau. The rainfall over the two main regions follows the sun, coming, broadly speaking, to a maximum after the sun is in the zenith and falling away as it recedes. Consequently there are two wet seasons over the Equatorial plateau corresponding closely with the equinoxes, although of the two, the spring rains are the more pronounced; while there is only one in Ethiopia coinciding with the summer solstice when the sun is at the zenith in the Northern Hemisphere.

The runoff from the two plateaus plays very different roles in the regimen of the modern Nile. The Equatorial plateau contributes a small but regular amount to the Nile in Egypt; and if this source of supply were to be cut off it can hardly be doubted, unless new sources come into operation, that the river Nile would run dry in the spring months. The rainfall on the Ethiopian highlands, on the other hand, produces the regular flood effect on which, until recent times, the whole agriculture of the country depended. Furthermore, the rich deposit brought down every year from the disintegration of the hill surfaces on which the rains fell was, until the building of the High Dam, both the architect and fertilizer of the flood plain which alone makes Egypt today a habitable country.

Taking Wadi Halfa (Sudanese Nubia) as a base representing what would have taken place in Egypt had there been no dams, barrages, or irrigation to interfere with the natural flow of the river, the levels of the river in the different months is shown in Figure 58 (after Hurst, 1944). Here the river begins to rise in June, and in the next two months it rises about 7 m reaching a maximum at the beginning of September, and then it falls more slowly than it rose to reach a minimum flow in May. Three streams, the Blue Nile, the White Nile, and Atbara, are responsible for the water supply of the Main Nile at Wadi Halfa, and the levels of these are shown on the diagram. Figure 59 (also after Hurst, 1944) shows the average discharge of the Main River at Aswan unaffected by reservoirs. If we take the maximum discharge occurring about September 8, we find that it totals 712 million m^3/day of which the White Nile contributes about 70 million m^3/day or 10% of the total, while the Blue Nile and the Atbara contribute about 485 and 157 million m^3/day or 68 and 22% of the total, respectively. The minimum discharge, however, is about 45 million m^3/day about May 10 and is made up of 37.5 million m^3/day from the White Nile (or about 83% of the total) and 7.5 million m^3/day from the Blue Nile (or about 17% of the total). The average discharge at Halfa is about 230 million m^3/day amounting to a total discharge of about 84 billion m^3/year of which close to 16% come from the Equatorial highlands. Discharges as low as 41 billion m^3/year and as high as 140 billion m^3/year are recorded depending on the amount of flood.

It is clear from the diagram that the largest quantity is contributed by the Blue Nile and the smallest by Atbara; but that at the low time of the year, from February to June, the White Nile is the more important source of supply. The Atbara contributes nothing from January to June. Figure 60 (also after Hurst, 1944) shows the average discharge of the Main Nile at Khartoum and the portions contributed by the Blue and White Niles. On the average 84% of the water of the Main Nile comes from Ethiopia and 16% from the Lake plateau of Central Africa.

Except for small and isolated tracts of land that have been recently reclaimed, the agricultural land of Egypt is the area that used to be annually flooded by the rise of the Nile. All land out of reach of the flood is virtual desert. The regular and stately precision of the rise and fall of the waters of the Nile, the composition and nature of the silt it carried, and the fact that the floods preceded the cool weather of Egypt all contributed in developing and perfecting the basin system of irrigation which remained, until very recent times, the only system of irrigation of Egyptian lands. Basin irrigation, as it was practiced in Egypt for thousands of years, was one of the most effective methods of utilizing the river. It can be started by the sparsest of

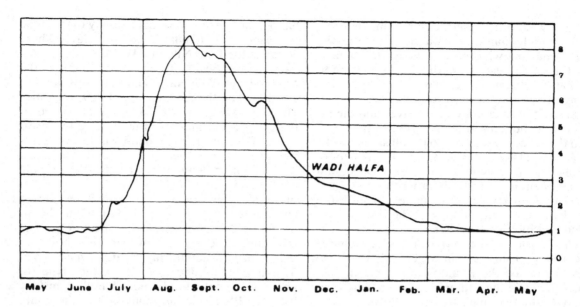

Figure 58. Water levels of the Nile at Wadi Halfa in
an average year (1931/1932). After Hurst (1944).

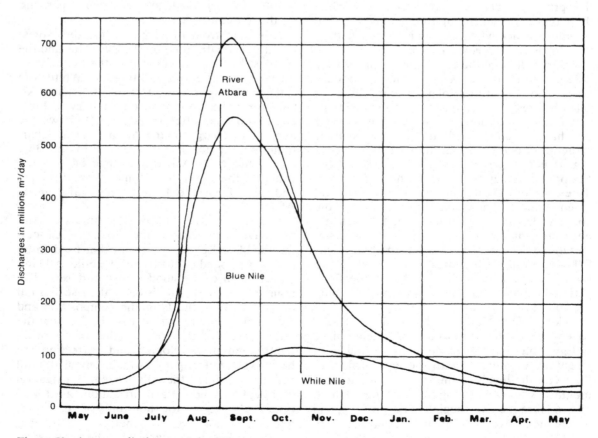

Figure 59. Average discharge of the Nile at Aswan,
unaffected by reservoir. After Hurst (1944).

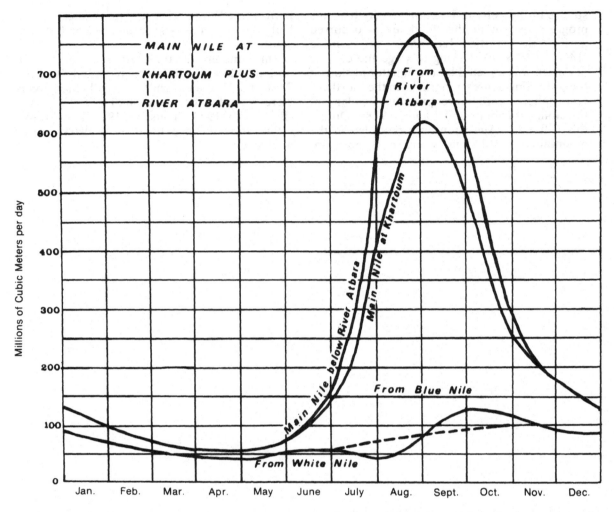

Figure 60. Average discharge of the Nile at Khartoum. After Hurst (1944).

populations and support in wealth a multitude of people, while the direct labor of cultivation is reduced to an absolute minimum.

Historical Records of the Nile Levels

The study of the records of the Nile levels which have been preserved for a long period (641 A.D. to present with short breaks) may help in understanding the nature of this regimen. In spite of the inaccuracies inherent in these records, they nevertheless form a valuable series which has been examined by numerous scholars. A valuable review of these records is given in Jarvis (1936) and Popper (1951). Hurst *et al.* (1966) analyze the available data and show that no periodicity seems to have governed the fluctuations of the floods of Egypt. Brooks (1949) shows that there is fairly good agreement between the flood level and low-water stage, although the fluctuations of the latter are the more violent. Both show a minimum at about 775 A.D., a maximum at about 870 A.D., a minimum at about 960 A.D., a maximum at 1110 A.D., and a double minimum at 1220 A.D. and 1300 A.D. It can be stated that starting from the fifteenth century the flood levels generally became progressively higher than during the earlier eight centuries. A study of the means of the maximum flood and low water levels per century, corrected for the progressive rise of the river bed due to deposition of silt, shows that the difference between the highest and lowest level became much larger starting from the fifteenth century (6.4 m) and continued to rise until it reached 7.3 m in the nineteenth century. Prior to this, however, the difference fluctuated between 5.6 and 6.1 m with an average of 6.25 m. While this rise in the level of the floods of the Nile could be attributed to the receding of the bifurcation of the delta distributaries from Cairo (where the measurements were made) with the result that the water surface rose in response to this retreat, it is noteworthy that this progressive rise in the flood levels occurred during the Little Ice Age of Europe (1430–1850 A.D.). During this age the climate of Africa was affected by the equatorial shift of the prevailing depression tracks in the Northern Hemisphere, with more prominent polar anticyclones (Gribbin and Lamb, 1978). During this Little Ice Age snow was recorded in the mountains of Ethiopia where it is now un-

known; the White Nile, fed from the equatorial belt, was low and rains weakened and shifted south.

Surviving historical records indicate that abnormally low floods were recorded between 2180 and 2130 B.C., while unusually high floods were recorded between 1840 and 1775 B.C. (Bell, 1970, 1971). Tousson (1923) and Brooks (1949) show that high floods were also recorded about 500 B.C.

CHAPTER III

Geological History and Concluding Remarks

The history of the Egyptian Nile Valley and delta is intimately connected with the geological evolution of northeastern Africa, the Eastern Mediterranean basin, and the Red Sea. Flowing today midway through the rainless wastes of the Sahara, the modern Nile gets its waters from sources outside Egypt, draining an area of more than 3,000,000 km². This enormous basin extending for 35° of latitude underwent great changes in recent geological history. These changes and the great climatic fluctuations of the past with their impact on world sea levels had their effect on the shape, regimen, and evolution of the river. In addition, the Eastern Mediterranean basin was subject to tremendous changes that converted it into an internal sea. It is fed by the overflow of waters from the Atlantic over the Gibraltar and Messina Straits. Eustatic lowering of the sea level would certainly sever its connection with the ocean and convert it into a basin with a stagnant anaerobic bottom such as occurred several times during the Pleistocene, (Thunnel et al., 1977), or even into a dry basin, as occurred during the late Miocene (Ryan et al., 1973a). The Red Sea also underwent enormous physical changes during the Neogene; its sea floor spread episodically during epochs as late as the Pliocene and Pleistocene. This must have affected the lands to its west, tilted some nearby regions, and forced their drainage along new lines. Furthermore, the Nubian massif which forms the bridge across which the Nile receives the waters of the equatorial and Ethiopian highlands witnessed great tectonic disturbances in the recent past. It was the site of great east-west transcurrent faults some of

which extend for hundreds of kilometers and have been active up to very recent time (Said, in Wendorf, 1964). The Sabaluka gorge which begins the long journey of the Nile waters into the waste of the Sahara is in effect a ring complex which lies along one of these great faults. The slightest tilting of the Nubian massif would indeed sever the connection of the Egyptian Nile from its sources and would cause it to decline or even to dry up.

During the past ten years numerous deep boreholes were drilled in the delta region and its offshore margins, and several hundred shallow wells were drilled in the Nile Valley and delta after the search for oil or water. These and the several thousands of kilometers of seismic work provide a clue to the history and origin of the Nile. The cross sections of the delta and valley (Figures 1, 2, and 3) are based on the information gained from the field mapping of the sediments of the valley, as well as on the data provided by the wells and seismic work. They give a glimpse of the paleogeography of the delta and the valley of the Nile prior to riverine sedimentation. However, more geophysical and geodetic work needs to be done in order to delineate the course of the ancient channel of the Nile, to determine the faults which cross it or run parallel to it, and to quantify the movements of these faults.

The study of fluviatile and other associated sediments of the Nile Valley shows that the river has undergone great changes since its downcutting in late Miocene time. This is to be expected in view of the complex and varied factors that have influenced its development. Five rivers succeeded one another in the valley.

These are from the oldest to the youngest: Eonile (late Miocene), Paleonile (late Pliocene), Proto-, Pre-, and Neoniles (Pleistocene). They were separated from one another by episodes in which the river declined, ceased to flow, or radically changed its regimen probably in response to tectonic activity and/or climatic changes.

The great climatic changes of the recent geological past had their impact on the land of Egypt itself which supplied most of the waters of the Nile during the early part of its history. Perhaps one of the most important results of the present work is the discovery that arid conditions did not set over the Sahara in general and Egypt in particular except during the Pleistocene. Previous to this epoch and during most of the Cenozoic, there is evidence that the climate in Egypt was wet. There was a good mat of vegetation, little surface denudation, and, during several epochs, moderate to intense chemical weathering.

Cenozoic rivers draining the elevated North African plateau have long been known, the best documented of which are the late Eocene, Oligocene, and early Miocene river systems which left behind great spreads of gravel and coarse sands as well as deltaic deposits in northern Egypt (Salem, 1976). These deposits were derived from the land of Egypt as well as from the lands beyond in Africa. Most of the fauna separated from these deposits have African affinities. The fluvial sediments of the Eocene, Oligocene, and Miocene rivers are spread over the Libyan plateau in the north. The mapping of large tracts of the limestone plateau in the middle latitudes of Egypt shows the presence of indurated fluvial gravel ridges lying unconformably over the Eocene surface some 200–400 m above the valley or the oases depressions (gravel fill on the Geological Map of Egypt, 1971). These ridges represent the remains of inverted wadis of post-Eocene pre-late Miocene drainage. Indeed the lowering of the surface of Egypt, the incision of the channel of the Nile and the formation of its canyon, and the beginning of the oases depressions all date back to the late Miocene in response to the lowered base level of the Mediterranean. It is indeed feasible to believe that the oases depressions were formed as a result of a drainage system which was graded to this new base level (Figure 32).

General aridity seems to have set over Egypt only during the Pleistocene. This arid regime was punctuated by episodes of increased local rainfall. During these short wet episodes of the Pleistocene accumulations of thick coarse gravels derived from local sources were deposited along the beds of many wadis. Two of these episodes, the Armant and Idfu, are the least documented but they seem to have been of relatively longer duration. They occurred around one and half million years and a million years ago, respectively. Three other maxima also occurred around 200,000 (Acheulian), 80,000 (Mousterian-Aterian), and 9000 y B.P. (final Paleolithic-Neolithic).

The effect of the Pleistocene pluvials can be seen in the southern reaches of Egypt where sediments and rich archeological materials are associated with the last pluvials (Wendorf et al., 1977). In contrast, northern Egypt seems to have undergone little geomorphic change in recent history and the effect of the pluvials is not evident. The presence of salt deposits probably as old as the late Miocene in the bottom of the Qattara depression (Said, 1979), the absence of drainage lines along the youthful escarpments of this depression and of playa sediments on its floor, and the scarcity of archeological sites of any significance in the deserts of northern Egypt all point to a stable landscape which was not influenced by the pluvials of the Pleistocene. In spite of the fact that the late Neogene was a wet period in Egypt and that the Pleistocene was interrupted by many pluvials, northern Egypt does not seem to have followed this pattern and seems to have been dry. The archeological studies of Hassan (personal communication) in northern Egypt and Wendorf et al. (1977) lead one to believe that Egypt was subjected to two different climatic models, the south having been affected by the northward migration of the Sudano-Sahelian savanna belt, and the north by the Mediterranean pre-Saharan steppe. Indeed the Pleistocene climatic changes can be related to changes in the general climatic setup of the recent. It can be assumed that a northward shift of the present-day climatic belts about 15° of the latitude they occupy now would bring an increase in precipitation in southern Egypt and would set a climate similar to that which must have prevailed during the Pleistocene "pluvials." The climatic pattern, however, during the pre-Pleistocene epochs of the Cenozoic must have been radically different, the explanation of which must assume a departure from the climatic pattern known today. It is noteworthy that the only probable resemblance to the prevailing Cenozoic climate during the Pleistocene is that which preceded the establishment of the Protonile regime. During that episode intensive chemical weathering must have prevailed since the gravels of this river system are characterized by the predominance of the more stable quartz and chert pebbles. This must indicate a vegetative cover and a wetter climate throughout the year to account for the weathering of the nonstable materials.

In spite of the fact that the Pleistocene pluvials were of relatively short duration, their impact on the geomorphological evolution of Egypt was immense especially in the southern parts of the country and in the Red Sea mountain range. The last of the great pluvials, the Korosko-Makhadma (Mousterian-Aterian), was responsible for the modern landscape of Egypt. As previously explained, the rolling topography of the modern Nile Valley was formed during this interval. In the southern reaches of the Western Desert great surfaces of erosion were developed and almost all the sediments of the earlier pluvials were washed away or were redistributed (Figure 32). In an earlier publication (Said, 1974) the author shows that save for a few relic old surfaces preserved because of very special conditions, the major pediplains of the south Western Desert of Egypt are of late Acheulian-Mousterian age and had been lowered only at a few localized areas. It is perhaps for this reason that little is known of the pre-late Acheulian archeology in Egypt, the sites of which have long been destroyed.

The depth of erosion to which southern Egypt was subjected during the last effective wet interval, the Mousterian-Aterian Korosko-Makhadma pluvial, can be seen from the degree to which the Abbassia gravels were lowered since their deposition during the earlier Acheulian wet interval. In Nubia the Abbassia gravels form inverted wadis which stand about 12 m higher than the surrounding plains. At the time of their deposition these ridges must have formed the thalwegs of valleys at least 30 m lower than the surrounding plains. It can be assumed, therefore, that during the Mousterian-Aterian pluvial the land was lowered by at least 50 m.

It is of interest to point out here that the Nile Valley and the deserts beyond have not undergone much geomorphological changes since the Korosko-Makhadma pluvial. Save for the filling of minor blowouts developing on the surfaces of the desert by playa deposits of the late Paleolithic-Neolithic pluvial, the pediplanation of the great sand sheets of the older alluviating rivers of southern Egypt, and the redistribution of their sand into the great dunes of the sand

sea, the deserts of Egypt show little geomorphological change during post-Mousterian-Aterian time.

On the other hand, northern Egypt, which remained arid almost throughout the Neogene and the Quaternary, shows very little geomorphological change since its elevation in post-middle Miocene time. The author (1979) advocates the idea that northern Egypt's landscape represents a fossil sea floor elevated and slightly modified since its emergence in post-middle Miocene time.

The Quaternary pluvials of southern Egypt are the following (Figure 61):

Armant Pluvial: Early Early Pleistocene pluvial depositing gravel and marl or sand beds of local derivation in many wadis. No master stream is noted in Egypt.

Idfu Pluvial: Late Early Pleistocene producing intense chemical weathering on the rejuvenated and elevated mountains of the Eastern Desert of Egypt and northern Sudan including most likely the Marra massif, Kordufan. A competent master stream, the Protonile, developed along a path 10–15 km to the west of the present Nile and having an elevation of 80–100 m above the present river. No archeological materials are associated with these deposits.

Abbassia Pluvial: Middle Pleistocene pluvial occurring during Acheulian time and depositing polygenetic gravels derived from the uncovered basement rocks of the Eastern Desert of Egypt and northern Sudan in channels which closely followed the path of the Prenile. Inverted wadis made up of these gravels in the West Nubian Desert indicate old drainage lines from this direction.

Korosko-Makhadma Pluvial: Late Pleistocene Pluvial occurring during Mousterian-Aterian time and forming slope wash or gravel–marl beds. The upper parts of the latter are exposed in the stumps of the major wadis of southern Egypt. The pluvial was responsible for the modern landscape of Egypt. The Nile Valley was excavated to greater depth than its present level and the surface of the surrounding areas was lowered by at least 50 m. Most erosion surfaces of the desert belong to this pluvial. Many of the channels of this pluvial were buried probably in response to to the eustatic lowering of sea level occurring at the beginning of the Wurm.

Dishna-Ineiba Pluvial: Holocene pluvial occur-

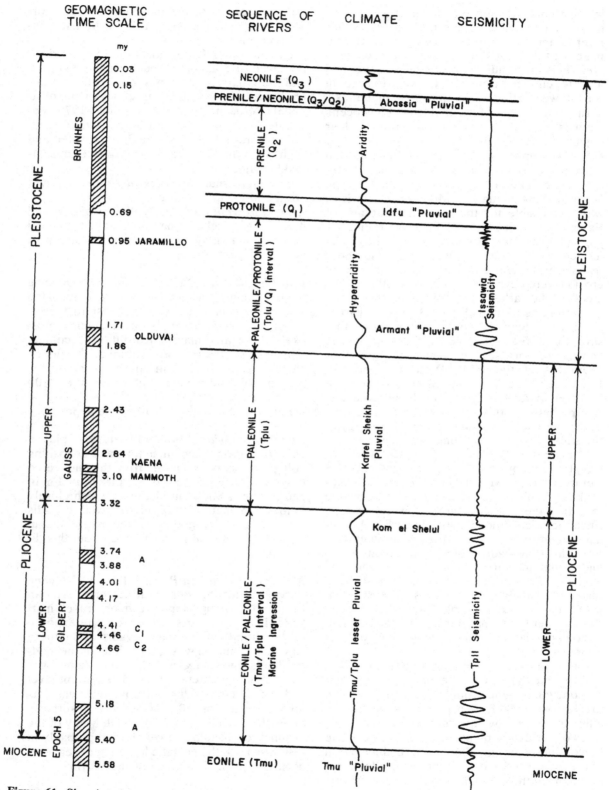

Figure 61. Showing dates of different Niles and accompanying climatic and seismic events, Vertical scale of upper part of time scale starting from 0.69 million B.P. is doubled. Geomagnetic time scale after Ryan (1973).

ring during late Paleolithic-Neolithic time. This pluvial was contemporaneous with the early aggradational phases of the younger silts of the Neonile and the preceding recession. In the southern Western Desert it filled ephemerally a few blowouts especially at the footslopes of the mountains. The playas thus formed were favored sites for human occupation (Wendorf *et al.*, 1977). This pluvial was not intense and continued to Predynastic or even Early Pharaonic times.

The Quaternary pluvials of southern Egypt are difficult to correlate with the world climatic events of that epoch. However, it can be stated that the Quaternary in Egypt was an epoch of great aridity punctuated by several pluvials none of which was sustained or vigorous enough to feed a Nile like the late Pliocene Paleonile or the late Miocene Eonile. On two occasions the Quaternary pluvials produced master streams, one of which was short in duration (the Protonile) and the other ephemeral in nature (the Abbassia). The Quaternary Niles in Egypt owed their existence to their connection with sources ouside Egypt and, in particular, with the Ethiopian highlands, a connection that took place during the middle Pleistocene and has continued on and off since then. Even though this connection was due to tectonic events, it is certain that the Nile could not have cut across the great wastes of the Sahara except during episodes of heavy rains on the Ethiopian highlands. Taken as a whole, it is noticeable that these episodes correspond to the glacial maxima of Europe. This was perhaps due to the effect of the glacials in accelerating the winds which swept over the African continent toward the Himalayas causing the great monsoons of East Africa and southwest Asia.

During the Mindel and Riss the Prenile had a great supply of water and sediment; and the breaking of the Neonile into Egypt coincided with the early Wurm. As has already been noted, the Little Ice Age of Europe was contemporaneous with a rise in the flood levels in Egypt. While the rains increased over the Ethiopian highlands during the glacials, the climate in Egypt became arid. Wendorf *et al.* (1977) demonstrate the aridity of the desert during the maximum Wurm glaciations. On the other hand, the pluvials of southern Egypt are correlated with the interglacials or the interstadials. The Acheulian pluvial (Abbassia) falls in the Riss/Wurm interglacial. During these intervals the climatic belts shifted to the north, and southern Egypt became part of the Savanna belt while northern Egypt became part of the Sahara.

III-1. THE MIDDLE MIOCENE (Tmm)

In middle Miocene time the northern part of the delta up to lat. 31° formed an embayment of the Mediterranean which was towered from the west and south sides by mountains of Eocene and Cretaceous rocks topped by Oligocene sands and basalt sheets. These mountains, which formed the South Delta block, must have been similar geomorphologically to the horizontally disposed table lands of northern Egypt and must have stood at a height of more than 1000 m above the floor of the middle Miocene sea. The deepest well drilled in the North Delta embayment did not penetrate through the middle Miocene; and the difference, therefore, between the top of the Eocene in the South Delta block (where it is reached at − 1701 m and − 1274 m in Mit Ghamr #1 and Shebin el-Kom #1 wells, respectively) and the North Delta embayment must have exceeded 3 km. In Damanhour well #1 the Upper Cretaceous chalk lies at a depth of −2568 m. In the Embayment itself the top of the middle Miocene lies at −4048 m in Kafr el-Sheikh well #1, at −3685 m in Sidi Salem well #1, and at −3759 m in the Baltim offshore well #1. These figures show the immensity of the downthrow of the complex of faults which bounded the South Delta block. According to the seismic data and drilling information available there seems to have been a zone bordering the northern edge of the block where a large number of step faults existed. These formed the hinge zone where the topography seems to have been subdued. In the Qantara #1 and San el-Hagar #1 wells, which lie in this zone, the top of the middle Miocene is encountered at −2150 m and −2389 m, respectively, more than 1000 m higher than in the North Delta embayment.

The faults of the hinge zone are of post-Eocene pre-middle Miocene age, and seem to be of the same age as the faults of the Cairo-Suez district (Said, 1962a) and those of the South Delta block. The downthrow of the South Delta block faults was smaller than that of the faults of the hinge zone. Basalt flows are closely associated with this faulting episode. Many of the wells drilled in this block show subsurface basalts in abnormal thicknesses. Thus, in Mit Ghamr well #1 the thickness of the basalt exceeds 325 m and occurs at a depth of −1117 m. The basalt also occurs at depth

−350 m in Abu Hammad well #1, −336 m in Wadi Natrun well #1, −90 m in Khatatba well #1, and −2 m in Abu Za'bal water well. The basalt also occurs on the surface along the southern edges of the delta. Indeed, one can speculate that the entire South Delta block was covered by one almost continuous gigantic sheet of basalt flow. The elevations given of the basalts indicate the nature of the old topography of this block which seems to have been affected by the orogenic stresses caused by the convergence of Eurasia and Africa (Galanopoulos, 1970; Wong *et al.*, 1971; Sonnenfeldt, 1974). It is interesting that, with the exception of a short episode of intense seismicity at the beginning of the Pleistocene, no further major movements have been recorded in the northern parts of Egypt since the middle Miocene.

The middle Miocene deposits of the North Delta embayment are thick and are made up of a solid shale unit (the Sidi Salem formation) which carries a rich open marine fauna. They were deposited in an open marine environment which was connected to the Gulf of Suez embayment by way of the shallow Suez isthmus. The Gulf of Suez has a similar facies and carries a similar fauna. The facies of the middle Miocene in the embayment contrasts with that of the Marmarica plateau to the west where shallow reefal limestones prevail (Said, 1962b). The South Delta block itself seems not to have

been overlapped by the middle Miocene sea although it was surrounded on almost all its sides by the shallow seas of that age. Figure 62 gives a simplified facies map of the middle Miocene sediments. The fact that the sediments are calcareous with little detrital material indicates that the climate was arid. The middle Miocene seems to have been a dry episode in an otherwise rainy period. There is no evidence that any sizable rivers were flowing into the Mediterranean during that age.

III-2. THE LATE MIOCENE (Tmu)

With the advent of the late Miocene there was a regression of the sea which ultimately resulted in the drying up of the North Delta embayment which had formed part of the Mediterranean. During the late Miocene the Mediterranean shrank to a series of inland salt lakes and finally to desiccation (Hsu, *et al.*, 1973). This remarkable discovery of Leg 13 of the Deep Sea Drilling Project which covered the Eastern Mediterranean basin finds supporting evidence in the drilling data from the Nile delta. Stromatolites, which indicate deposition in a shallow environment, are found at great depths. In Sidi Salem #1 they occur at depth 3460 m. In addition, anhydrite beds are found in many of the wells drilled in the North Delta embayment (see

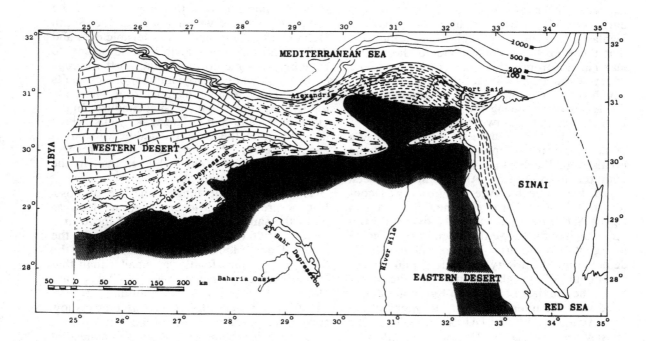

Figure 62. Egypt during middle Miocene time.

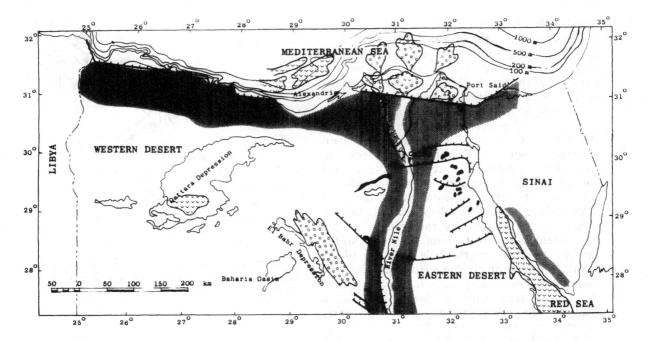

Figure 63. Egypt during late Miocene time. Remains are also shown of early Miocene fluviatile sediments of river leading to Moghra, eastern tip of Qattara depression.

Table II-1 for their distribution and depth). These anhydrite beds are correlated with the strong acoustic M reflector now known to result from an evaporite layer found at shallower depths throughout the Mediterranean below the sea bottom and closely simulating the bottom topography. These beds form a magnificent stratigraphic marker terminating the Miocene epoch; the end of the Miocene, according to the most recent research, falls around 5.4 million years B.P. (Ryan, 1973; Cita, 1973a,b; and discussion by Berggren, 1969, and Berggren and van Couvering, 1974).

With this new geomorphology and lowered sea level, erosion became extremely active; and it was during this time that the valley of the Nile started being formed by a river which is here termed the Eonile. As the sea dried up the collected water vapor of the desiccated Mediterranean was moved by the northerly winds to the elevated ranges of the Eastern Desert of Egypt causing abundant rains all the year round. The wadis through which these rains were channeled incised deep gorges in these mountain ranges which were still topped by a Cretaceous and Eocene cover which has since then disappeared. The wadis seem to have continued to cut their channels in the underlying

folded Precambrian rocks (which build up the modern Red Sea range) forming a superimposed drainage which ultimately found its way to the everdeepening Eonile. The course followed by this river was certainly determined by the tectonic framework of Egypt at the time. It was the first time in the history of the drainage of the elevated lands of Egypt that the river followed a defined and deep gorge which distinguishes it from the alluviating and shallow rivers of the pre-late Miocene drainage systems which preceded the Nile system. The river immediately preceding the Eonile (early Miocene) followed a northwesterly course from Assiut, traversing the Western Desert and debouching at Moghra oasis at the eastern tip of the Qattara depression (Figures 32 and 63). The deflection of the course of the Eonile to a more northerly course at Assiut was initiated by a tilting movement at the critical Minia-Assiut stretch. This new course followed a subsequent drainage system which developed from the consequent drainage of the newly elevated Red Sea range. The headward erosion of these subsequent streams seems to have determined the course of the Eonile in this stretch.

The Eonile formed a deep canyon the bottom of which reached depths ranging from −170 m at Aswan (Chumakov, 1967) to −800 m in Assiut and to more than 2500 m in the recently discovered channel to the north of Cairo (see Appendix C) and to even greater depths in the North Delta embayment (over 4000 m in the

Kafr el-Sheikh area). The river seems to have cut its channel in the elevated north Egyptian plateau passing through the South Delta block and cascading over the hinge zone until it fanned into the North Delta embayment depositing its load in a series of coalescing fans.

The shape and dimensions of the Eonile canyon must have resembled to a great extent the Grand Canyon of the Colorado River, Arizona, although the Eonile canyon seems to have been longer and deeper. The Eonile canyon had approximately the same width as the Colorado canyon. Both rivers cut their paths in bare horizontally disposed sedimentary strata varying in lithology and color. Both are antecedent rivers, the Colorado by virtue of the rise in landscape and the Nile by virtue of the continuous lowering of the base level. The Nile canyon, however, was cut in a very short time during the late Miocene, while the Grand Canyon has been forming since the early Miocene up to the present. The Colorado River seems to be a small river in comparison with the Eonile although it could well compare with the modern Nile. Both the Colorado and the modern Nile cross large stretches of desert country but receive sufficient waters from sources beyond to carry them through successfully. Both carry equal loads of sediment in the range of 300,000 tons/day. However, the Eonile must have been an extremely vigorous river dumping an estimated 70,000 km³ of sediment in less than three million years or about 20% of the total amount of sediment in the delta cone. The gradient of the Eonile averaged about 1:400 from Aswan to Cairo (1:250 in the stretch from Assiut to Cairo) as compared to that of the modern Nile which is 1:13,000 from Aswan to Cairo. The following table gives a comparison between the Grand Canyon and the Eonile canyon.

	Grand Canyon, Arizona	Eonile canyon
Width	10–20 km	10–20 km
Length	320 km	1300 km
Depth	2080 m	2500 m
Gradient	1:625	1:400

Most of the sediment transported by the Eonile came from the Cretaceous and the Eocene as can be attested from the lithology of the rolled pebbles found in the Eonile sedi-

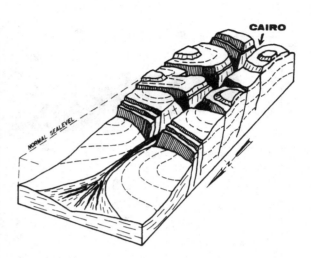

Figure 64. Block diagram of the Eonile Canyon from the area of present day Cairo to the desiccated Mediterranean of late Miocene time. (Drawing by Felix P. Bentz)

ments. However, in the South Delta block the Eonile canyon cut through a thick basalt sheet and Oligocene sand and gravel beds before it cut the Eocene, Cretaceous, and finally the Jurassic rocks which most likely made the bottom of the canyon. Figure 63 sketches the paleogeographic reconstruction of Egypt in late Miocene time, and Figure 64 is a block diagram of the Eonile canyon drawn by Dr. Felix Bentz.

III-3. THE EARLY PLIOCENE (Tpll)

The advent of the Pliocene in the Mediterranean region was marked by an advance of the sea over the desiccated Mediterranean basin through an inflow from the Atlantic over the Gibralter Straits (Hsu and Cita, 1973). The rising sea level of the early Pliocene brought the Mediterranean into the Nile Valley depression transforming it into a narrow long gulf reaching as far south as Aswan in Upper Egypt (Figure 65). The distribution of the early Pliocene deposits in northern Egypt shows that, in addition to the drowned delta and the Nile canyon, the Pliocene sea covered large tracts of the lands around the modern delta especially along its western edges (Figure 8), but did not overlap except very small fringes of the present day coast of Egypt. The Mediterranean–Red Sea seaway effective during the Miocene was severed and the coastal regions of the Mediterranean were land. The South Delta block, already in existence since the middle Miocene, was

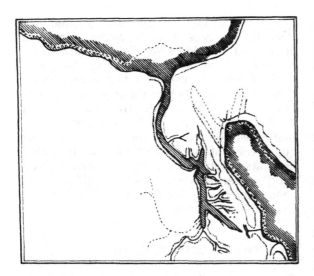

Figure 65. Egypt during early Pliocene time.

overlapped by the sea after it had been appreciably lowered down by erosion.

The deposits of the deeper parts of the drowned canyon are in the form of sands (18–59%) and montmorillonitic clays with thin lenses of fine-grained polymictic sands and sandy loams rich in authigenic minerals: glauconite, pyrite, and siderite. Toward the peripheries of the canyon the deposits are in the form of sandy limestones, marls and coquinal beds carrying marine fossils; the beds abut against the bounding cliffs of the canyon. The fossiliferous beds of this episode (best exposed in Kom el-Shelul, south of Gizeh Pyramids) rest on the slipped masses of Eocene or, more frequently, on the Eocene bedrock itself. An intervening band of conglomerate or breccia up to 3 m in thickness separates the marine Pliocene from the underlying bedrock. Little (1935) maps the Pliocene beds of this shrunken gulf in order to reconstruct its outline. According to Little, the average width of the gulf was 12 km but the arms of the sea extended inland to some distance on either side (see accompanying geological map). The longest inlet was on the western side and had a length of 30 km; on the eastern side the longest was only 11 km. Near Dahshour, the distance from the head of one inlet across the gulf to the head of the inlet on the opposite side was 44 km. The maximum height at which the shelly marine beds are found is 125 m above sea level.

The Kom el-Shelul formation shows that carbonate deposition prevailed during the early Pliocene. This may be taken as evidence that it was deposited during a cool episode. Arrhenius

(1952) points out that cyclic carbonate sedimentation found in the Pacific cores he studied is related to climatic fluctuations. Generally from about 5.4 million years ago onward, episodes of inferred warm climates appear to correlate with carbonate minimum (see also discussion by Cita and Ryan, 1973; Hays et al., 1969). Supporting evidence for this view comes from the work of Said (1955) who studies the foraminifera separated from the Kom el-Shelul formation and finds out that it is boreal in habitat. Following the then current ideas Said had suggested a lower Pleistocene age for this formation. However, recent work on the stratotypes of the Mediterranean Pliocene and Pleistocene epochs, as well as the cores raised from the Red Sea, shows that the Pliocene/Pleistocene boundary in this region occurs at the Olduvai magnetic event of the Matuyama epoch at an interpolated age of 1.85 million years, and that the Pliocene epoch includes several cold episodes. It is plausible, if one accepts the time scale adopted by the workers on the Mediterranean cores and accepted by others (see discussion by Berggren, 1969, 1971, 1972; and Berggren et al. 1967; Berggren and van Couvering, 1974), to postulate the age of the Kom el-Shelul formation as falling in Ciaranfi and Cita's (1973) climatic episode "red." On the climatic curve this episode is characterized from later to earlier first by cooling, then by warming, and finally by another cooling.

The marine sediments of the early Pliocene filled about one-third of the depth of the ancient Eonile canyon. The thickness of the sediments on the south Delta block away from the gorge and in the Nile valley was in the range of 300 m, while in the North Delta embayment a thickness of over 1000 m is recorded. The amount of sediment deposited in the delta cone during this age amounted to about 110,000 km³.

III-4. THE LATE PLIOCENE (Tplu)

Reference has already been made to the effect of fresh water on the marine sediments of the early Pliocene gulf of the Nile Valley; but this effect does not seem to have changed the essentially marine character of the gulf except later in time, perhaps at the start of the late Pliocene fixed by many workers at 3.3 million years ago at the Gilbert-Gauss magnetic interval and marked by a noticeable deterioration of the climate (Hays et al., 1969). This cool climatic episode, designated "brown" by Giaranfi and Cita (1973) in the Mediterranean, corresponds well with the base of the Nebraskan (as recorded in the Gulf of Mexico), the onset of ice-drifted detritus in the Northern Atlantic, and the Middle Villafranchian continental fauna. This cold interval is considered by many authors as the beginning of the Quaternary (Blanc, 1955; Lamb, 1969). However, the adoption of such a boundary is inconsistent with the boundary as defined at the INQUA Congress (1955) and would imply that the faunal assemblages of the classical section of Castellarquato should be excluded from the Pliocene. Since this section has had worldwide use in biostratigraphic correlations, the consensus is to keep it in the Pliocene.

The effect of the late Pliocene river, here termed the Paleonile, was to convert the earlier Nile Valley gulf into a veritable channel of a river which opened up into the embayment to the north influencing the salinity of its waters; the faunas started to include elements of brackish water habitat. It is on the basis of this change of faunal assemblages that the present author divides the column of Pliocene sediments recored in the delta deep boreholes. Figure 16 gives the structure contour of the base of the Paleonile sediments, while Figure 17 gives the isopachs of these sediments. Both figures show that both the Eonile canyon and the Paleonile channel follow, by remarkable coincidence, one of the major seismic belts of the delta as delineated by Gergawi and Khashab (1968a) from a study of the earthquakes in Egypt occurring during the years 1951–1966. At an early date of depositon the sediments overflowed the channel and started building up a delta that fanned out as far west as Wadi el-Natrun, where a mixed fluviomarine fauna is found, and then as far north as the edge of the delta cone, as recently delimited by the oceanographic vessels which cruised the Mediterranean (Emery et al., 1966). The Paleonile sediments make about 20% of the section of riverine deposits in the valley and delta. An estimate of their volume is 70,000 km^3. By the end of the Paleonile sedimentation the Eonile canyon was completely filled up and the delta surface became more or less even with a northward slope.

The lithology and mineral composition of the Paleonile sediments are uniform, and their source must be sought in areas receiving sufficient precipitation. The fact that the Paleonile sediments are completely lacking in coarse detrital materials but are rich in montmorillonitic clays and organic matter and that the iron is exclusively present in the ferrous form (Chuma-

kov, 1967) formed under reducing conditions indicates sources with effective vegetation cover and considerable moisture distributed fairly evenly over the year. The absence of Central African freshwater faunal elements in these beds suggests that these sediments must have been largely provided from Egypt, probably from highlands in the Eastern Desert. This points also to extremely wet climates over this country. Proof that the Paleonile received great amounts of detritus from the numerous wadis of the Egyptian Red Sea hills is provided by the sediments fringing many of the wadis that drain today into the valley.

III-5. THE PLEISTOCENE (Q)

The advent of the Pleistocene epoch in Egypt was marked by dramatic events the most important of which, relative to their impact on the history of the Nile, were those relating to tectonism and climate. The beginning of the Pleistocene epoch in Egypt was marked by an episode of high seismicity recognizable in the thick talus breccias which accumulated along the piedmont slopes of the bounding cliffs of the valley and the wadis that drained into it. This episode coincided with the intense episode of tectonic activity which set over the Red Sea area when sea-floor spreading resulted in the formation of its axial zone (Ross and Schlee, 1973). It also coincides with the recent tectonism postulated by Hsu et al. (1973) to have occurred in the Eastern Mediterranean basin along the outer part of the Nile cone.

The other important dramatic change which took place at the advent of the Pleistocene was the great climatic change that occurred over Egypt during that time, bringing to it a pattern of aridity that set the tone of the climate prevailing in Egypt, with minor fluctuations, throughout the Pleistocene.

In the following paragraphs a resume is given of the Pleistocene geological history of the Nile Valley during the early (Q$_1$), middle (Q$_2$), and late (Q$_3$) Pleistocene. The tripartite division of the Egyptian Pleistocene compares with the classification recently proposed for Europe (Kukla, 1977).

The Early Pleistocene (Q$_1$)
This interval covers the time span which elapsed from the advent of the Pleistocene fixed at 1.85 million B.P. (Ryan, 1973) to ca. 700,000 B.P., the time of the first great Pleistocene cooling marking the advent of continental

glaciation in temperate latitudes (Hays and Berggren, 1970; Ericson *et al.*, 1968; Evans, 1972). This long time extending for more than one million years covers the Paleonile/Protonile as well as the Protonile intervals. The Paleonile/Protonile interval was characterized by the conversion of Egypt into a veritable desert. Not only did aridity set over the country, but the Paleonile itself stopped flowing into Egypt. It was an interval marked also by intense tectonism leading to the accumulation of thick breccias at the footslopes of the cliffs of the valley and the wadis which drain into it. The best exposures of these talus breccias is in the neighborhood of Akhmim.

In the early phases of this interval a short pluvial seems to have occurred leading to the accumulation of more than 40-m thick beds of locally derived coarse detritus (Armant formation). This pluvial seems to have been terminated by the deposition of thick travertines either on the slopes of the bounding cliffs of the valley, as in Naga Hammadi, or on horizontal surfaces of the beds of inland and peripheral ponds, as in Issawia. Most of the travertines hitherto recorded are associated with the limestone cliffs of the middle latitudes of the Nile Valley. In places, during the end phase of the Armant pluvial, the drying up of the percolating waters produced tufaceous limestones which filled the interstices of rocks cementing them (such as the well-known red breccia quarried for ornamental use since the time of ancient Egypt).

The great aridity which set over Egypt during this interval started the process of wind deflation that is still going on to this day contributing to the formation of the modern landscape of Egypt. The great desert depressions and the great surfaces of erosion known in the plains and table lands of Egypt, both to the east and west of the Nile, were affected by this interval of aridity. This, together with the erosive action of the rains of the different pluvials, lowered the desert surface by several hundred meters. There is evidence in the south Western Desert (Said, 1974) that the difference in elevation between a relic "? Neogene" surface and the modern surface exceeds 200 m.

Toward the end of the early Pleistocene a highly competent river, the Protonile, seems to have occupied the Nile Valley. It carried cobble and gravel-sized sediment made up mainly of quartz and quartzite probably formed on a deeply leached terrain. This is one of the most characteristic features of these early Pleistocene gravels and distinguishes them from

later gravels, for it indicates the derivation of the Protonile sediments from terrains with less continentality. The source of this river is unknown, but it certainly derived part of its waters from areas outside Egypt; the Protonile gravels are known all along the Egyptian Nubian Nile and perhaps also in the Khartoum area (Arkell, 1949). Other sources of the river may have been within the land of Egypt, for gravels made up of quartz and quartzite are also recorded from the higher terraces of Wadi Allaqi (the so-called 32- and 21-m terraces of this wadi recognized by Said and Issawi, 1964). These are found to be different in the composition of their gravel from those of the lower terraces of the same wadi which are of polygenetic composition. In Wadi Gabgaba, Eastern Nubia, there are gravels of similar composition. The significance of these siliceous gravels is that they indicate most probably that the Protonile derived most of its waters from the same areas as the Paleonile. These areas were subjected to intense chemical disintegration during the wet epoch of the late Pliocene and were stripped of their vegetation cover; and their finer disintegration products were deflated leaving behind bare mountainous areas covered with these siliceous gravels only. No Paleonile sediments are found in wadis Allaqi or Gabgaba (such as those in the Kharit-Garara wadis); but this may be due to the fact that both lie in the uparched lands of Nubia which were elevated during the early Pleistocene and redissected in later time so that the older river sediments, if they ever existed, were eroded away. Both wadis are great trunk wadis of the Nile; Allaqi alone, which reaches a length of more than 350 km, drains an area of close to 45,000 km^2. Both wadis derive their waters from the same sources as the Kharit-Garara complex.

The duration of the Protonile must have been short; its tributaries do not seem to have had time to cut their way in the underlying disintegrated but little leached terrain; the Protonile gravels are relatively thin and do not show any change with time in their lithology.

In conclusion, it can be stated that the Protonile covered a short period of time probably contemporaneous with the earliest glaciations of the Pleistocene. The climate must have been wet, similar to that of the Pliocene, but of too short a duration to produce effective vegetation cover or intense chemical weathering. Soils on the surface of the Protonile are red-brick in color, but these seem to have developed in later pluvials. The sources of the Protonile seem to have been the same as those of the Paleonile;

but the Protonile was a more competent river (as it followed an interval of intense tectonism and rejuvenation of relief) deriving waters from bare mountains with surfaces showing the effects of an earlier intense interval when chemical weathering prevailed.

The Middle Pleistocene (Q_2)

No sediments have been recognized so far which characterize the interval which separated the Protonile from the succeeding Prenile; but this must have been an interval of important events in regions outside Egypt. These events led to the severing of the connections with the drainage lines that had so far supplied the three old rivers and brought in new connections with the Ethiopian highlands. Since the breaking through of the Prenile into the land of Egypt, these new connections have continued to form the main sources of supply for the Nile up to the present time. The interval, therefore, must have been characterized not only by an arid climate, but also by tectonic disturbances in the Red Sea which paved the way for the capture of the Atbara and Blue Nile sources. The Prenile was a vigorous and competent river. It brought to the land of Egypt large amounts of sediments, mainly sands estimated to be in the range of 100,000 km³. The mineralogy of these sediments is radically different from that of the sediments of the earlier river inasmuch as they include an influx of new minerals such as those characterizing the modern Nile (Shukri, 1950). Even though the heavy mineral suite of the Prenile sediments is different from that of the modern Nile, and particularly in the abundance of epidotes relative to pyroxenes, it includes, nevertheless, many common mineral species to warrant the conclusion that both had a common provenance. The nearest model of drainage that could explain this change of mineral composition is a larger influx of waters from the Atbara relative to the Blue Nile.

The Prenile sediments underlie the Abassia gravels which are of Acheulian age believed to have centered around 200,000 B.P. The Prenile, therefore, must have terminated about this date. The beginning of the Prenile is difficult to determine, but it could be set, on stratigraphic grounds, at the beginning of the middle Pleistocene. If this is accepted then the Prenile must have been flowing in Egypt during the Riss and Mindel glaciations for a duration of close to half a million years.

The Prenile was a competent river carrying sands which were deposited on its large flood plains and expanding delta, both of which exceeded in extent those of the modern Nile.

These sediments also occur in the Faiyum depression, as the Prenile during its last aggradational phase reached the Lahun gap and had access to the depression forming a lake of large dimensions which reached an absolute elevation of 44 m (the elevation of the river today at the gap is 27 m). The sediments of the Prenile are described elsewhere in this book. They deserve a more detailed study; for they seem to have had a complex history perhaps with numerous aggradations and recessional phases.

The path of the river was a channel that occupied a more westerly course to that of the modern Nile, but eastward to that of the preceding Protonile. No Prenile sediments have been recorded so far from Nubia which seems to have been undergoing erosion throughout the duration of the Prenile. The episodic uplifting of Nubia about 200,000 years ago probably affected the flow of the Prenile into Egypt and reduced it to very low levels.

The middle Pleistocene terminated with a pluvial, the Abbassia, which was characterized by intense rains over Egypt resulting in the accumulation of thick locally derived gravel deposits which rest unconformably over the eroded and stabilized surfaces of the Prenile and earlier sediments. The gravels are polygenetic and contain abundant crystalline rocks and feldspathic sands derived from deeply disintegrated but little leached terrain. They represent the first generation of gravels derived from the uncovered basement rocks of the Eastern Desert of Egypt. They are very similar to the massive gravels which accumulate today in the wadis which drain the Eastern Desert of Egypt after the torrential winter rains.

The Abbassia gravels represent one of the most conspicuous stratigraphic horizons in the entire Nile sequence. They are rich in archeological material of late Acheulian tradition. The distribution of the Abbassia gravels indicates derivation from local sources particularly the Red Sea area. The composition, thickness and lithology of the gravels indicate a short period when winter-season cyclonic cloud bursts were intense and much more frequent than at present. The Abbassia pluvial corresponds probably to the Riss/Wurm interval.

During the Abbassia pluvial the bed of the Prenile sediments was cut to lower levels, and during the succeeding intervals the modern valley assumed its final shape.

The Late Pleistocene (Q_3)

The late Pleistocene of the Nile Valley is represented by the deposits of the Neonile which broke into Egypt some time in the the earlier part of this age and also by the deposits which accumulated during the recessional phases of this river. The Neonile was a very humble successor of the Prenile; its volume was one-fifth of that of the Prenile, and it was interrupted by long episodes of recession and minor flow. It is almost certain that the breaking of the Neonile in Egypt occurred as a result of the flow of the Blue Nile and Atbara across the elevated Nubian massif by way of a series of cataracts. Throughout its history the Neonile in this massif has been continuously lowering its course at a rate of 1 m/1000 years; the older sediments of the Neonile appear as benches 27 m higher than the river's flood plain and the younger sediments appear at successively lower elevations. The oldest or α Neonile was followed by a long period of recession in which the waters of the Nile almost dried up or were lowered down to levels which cannot be determined. During this recessional interval a major pluvial occurred (Mousterian-Aterian) and great geomorphological changes took place over the valley. This interval was followed by three aggradations (the Masmas-Ballana, Sahaba-Darau, and Arkin) which were interrupted by minor recessional episodes (the Deir el-Fakhuri and Dishna-Ineiba). The breaking of the last three rivers (β, γ, and δ Neoniles) into Egypt coincides with the age of the deeps in which hot brines have been recently discovered in the Red Sea basin (Hachett and Bischoff, 1973). The fissuring of these deeps seems to have marked an episode of tectonic activity which affected the Ethiopian highlands and the direction and intensity of the flow of the waters that fell on them.

In the Egyptian valley itself, the Neonile aggraded its beds and deposited its sediments over the eroded and uneven surface of the flood plains of the Prenile molded during the Abbassia interval. The composite thickness of the sediments deposited by the Neonile represents less than 1% of the total sediment deposited by the earlier rivers which occupied the valley. The earlier aggradational episodes had higher floods than are known today, and this trend continued after the Sahaba-Darau aggradation which seems to have had the highest floods known in the history of the Neonile. Judging from the elevation assumed by the Sahaba-Darau and the relative coarseness of its sediments, which include thin pebble beds and sands, the floods probably averaged 140 billion m³/year, a volume that has seldom been reached by the modern Nile. Posterior to this aggradation and after a short recession (Dishna-Ineiba) the Neonile at 10,000 B.P. assumed a

regimen and gradient very similar to those of the modern Nile. The floods seem to have averaged over the years, as in modern times, about 80 billion m³/year. The episodes of recession which interrupted the intervals of aggradation were episodes when the floods were low, probably in the range of 40 billion m³/year, a volume which was recorded by the modern Nile during the years of pestilence known in the history of Egypt. The Holocene Nile seems to have aggraded its bed in response to an advancing sea; the delta must have had, at the beginning of the Holocene, a coast that probably lay 50 km to the north of the modern coast. The longshore currents, which must also have established a regimen like that of the present, transported the detritus of the river toward the east building up numerous barrier beaches in the Gulf of Pelusium. These beaches were gradually incorporated in the land causing the coast of Pelusium to advance at the rate of 5 m/year.

During the aggradational episodes the climate was very similar to the climate prevailing over Egypt today. Save for very occasional cloudbursts, the wadis were not active. Wind-blown sands interfinger the deposits of the Nile, especially along the west bank of the river where an enormous supply of sand and great fetch produced dunes of sizable dimensions. Dunes interfinger the Masmas-Ballana aggradational deposits (Wendorf and Said, 1967; Wendorf and Schild, 1976) and the Sahaba aggradational deposits in a most spectacular manner along the west bank of the Minia-Mallawi stretch of the Nile. These form the famous Khefoug landscape which makes a distinct mappable rock unit that appears on the accompanying geological maps of the valley and delta. It is interesting to note that the Prenile sediments in this same stretch have thick interfingering dunes. The Khefoug deserves special attention, for this is the only stretch on the banks of the river where sand dunes form the major component of the exposed Nile terrace. Thin laminae of Nile silts interfinger the deposit which is otherwise made up of dune sand which is now stabilized.

The path of the Neonile lay to the east of that of the Prenile. However, the river shifted its course toward the west during the Holocene. Thus, in many parts of the Nile the Sahaba-Darau channel, which silted up during the Dishna recession, was abandoned by the succeeding Arkin River to new westward channels which in most cases lay in hard rock. The classical example of this is the abandoned Aswan-Shallal channel which seems to have formed the path of the river during earlier times and up to Sahaba time (a drillhole in the channel repeats bed by bed the section of the Aswan High Dam made famous by the descriptions of Chumakov, 1967). The Arkin River abandoned this well-established channel and followed the difficult path across the granitic mass of Aswan interrupted today by the well-known Aswan cataract. Another example of this tendency of the river to shift its bed to the west occurs in the Silsila region to the north of Kom Ombo where again, in Arkin time, the river abandoned its established channel to follow the difficult presentday path through the siliceous sandstones of Gebel Silsila, cutting its way through a gorge. Less spectacular examples are present in the Armenna region of Nubia, now under water, but shown on the map of Butzer and Hansen (1968) where again the Holocene river abandoned its old channel and followed a westward course through the Nubia sandstone. In Abnub, Assiut province, a similar phenomenon is noted.

Thus, the Neonile which followed a course that lay to the east of its predecessor, washing in many places the Eocene eastern cliffs that border the valley in its middle latitudes and by-passing the ridge of Eocene limestone that lies underneath the surface at shallow depth within this reach (section 6-6', Figure 3), changed this trend in Holocene time and started to move toward the west in several places all along the Nile. This trend seems to have been in response to movements at the inception of the Holocene which tilted the older silted beds of the Nile in certain localized areas. The silting up of the distributaries of the delta started only after this early Holocene movement which also seems to have affected the position of the branching of the delta distributaries which is known to have shifted northward during the Holocene.

Studies on the mineral composition of the Nile sediments are given by Shukri (1950), el-Gabaly and Khadr (1962), Hamdi (1967), Butzer and Hansen (1968), and Hassan (1974). Venkatarathnam and Ryan (1971) study the clay mineralogy of surface sediments of the Nile cone in the Mediterranean and show that a distinct clay mineral assemblage with high amouonts of well-crystallized smectite (less than 50%) and 15–25% kaolinite occurs in these surface sediments. These two minerals characterize the Neonile sediments. Their distribution in the Mediterranean, like the sands of the modern river, appears to have been related to easterly moving surface currents which still form part of the counterclockwise gyre present in the modern Eastern Mediterranean. These currents seem to have continued for the duration of the Holocene contributing to the building of the advanced coast line of the Gulf of Pelusium.

It is noteworthy that the reconstruction of the history of the Neonile fits many facts and observations that have been recently revealed through the work of many scholars. The study of the stratigraphy of the Nile cone late Quaternary sediments and their dispersal (Maldonado and Stanley, 1976; Stanley and Maldonado, 1977) shows that the earliest of the modern Nile sediments reached the cone at 38,000 B.P. Earlier periods, of which there is reasonable chronological control (i.e.,58,000–38,000 B.P.), had minimal rates of sedimentation. Between 38,000 and 28,000 B.P. terrigenous sediments became abundant especially around the Rosetta branch. Higher rates of sedimentation occurred between 28,000 and 23,000 B.P. when they declined until 17,000 B.P.

This curve of sedimentation (Figure 66) throws light on the history of the river and the paleoclimatic model which could explain the fluctuation of the discharge of the river as well as the building of the Nile delta and cone. Vita-Finzi's work (1972) gives a generalized curve of the fluvial supply of sediments to the Mediterranean; they are shown to belong to two fills, one fill between 50,000 and 10,000 B.P. and the other between 2000 and 300 B.P. Interestingly, the dates given by Stanley and Maldonado coincide, to a large extent, with major events of the Neonile as conceived in this work. The period of low sedimentation rate prior to 38,000 B.P. coincides with the Korosko-Makhadma interval when the river was very low or even wanting. The period of overall highest sedimentation occurred between about 28,000 and 17,000 B.P. with lesser rates between 23,000 and 17,000 B.P. This coincides exactly with the span of the Masmas-Ballana episode when the waters are known to have overflowed the banks of the river carrying large amounts of sediment. This would contradict Fairbridge's thesis (1962, 1967) that the Nile aggradation and widespread silt deposition imply reduced competence and a smaller discharge of the river. It seems to substantiate the thesis held by de Heinzelin (1968) and Wendorf, *et al.* (1970 a,c) who equate the Nile's aggradation with conditions of maximum flow.

It is also interesting that the Deir el-Fakhuri recession seems to correlate with conditions of lesser discharge prevailing everywhere in Africa during that time. The water level was

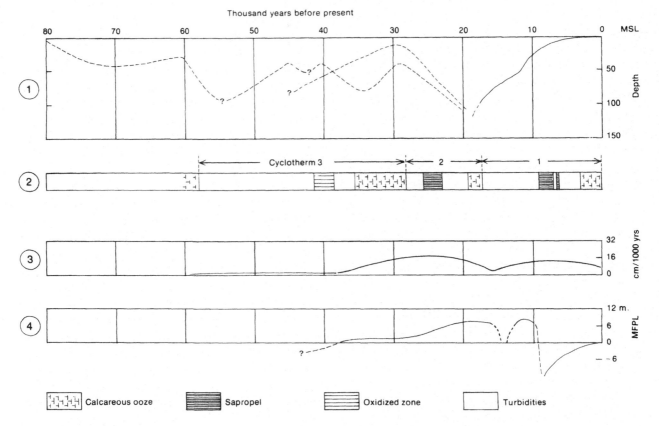

Figure 66. Correlation between rates of sedimentation in the Nile cone (3), the level of the river Nile in Upper Egypt (4), the eustatic sea level curve (generalized) (1) and Nile cone stratigraphic type core (2). Part of the data after Stanley and Maldonado (1977). MFPL, mean flood plain level.

low in the White Nile previous to 12,000 B.P. and sand dunes were being formed (Williams and Adamson, 1974; Saad and Sami, 1976). Geochemical and palynological evidence indicates that during the same period Lake Victoria had no outlet and dry forest grew on the Rowenzori (Kendall, 1969). The levels of lakes Chad, Rudolf, and Nakru were also very low at that time, and dunes were active in the now vegetated parts of Nigeria (Servant, 1973; Butzer *et al.*, 1972).

About 12,000 B.P. Lake Victoria overflowed into the Nile, the level of the White Nile rose and the Sahaba-Darau aggradation began in Egypt. Figure 66 shows the correlation between the rates of sedimentation in the Nile cone, the level of the Neonile in Upper Egypt and the eustatic sea level curve.

APPENDIX A   Tables of Boundaries of Rock and Time-Rock
                    Units in Deep and Shallow Boreholes and in
                    Measured Surface Sections

**Table A-1. Boundaries of Time-Rock Units in Deep Boreholes
Drilled in Nile Delta, Egypt**[a]

No.	Well	Lat. N	Long. E	Elev. (m)	Depth (m)	Q_3
1	el-Tabia #1	31°19′05″	30°03′25″	+ 2.5	2395	—
2	Kafr el-Dawar #IX	31°02′28″	30°05′24″	+ 0.2	2652	0.0
3	Hosh Isa #IX	30°55′26″	30°17′13″	+ 0.5	2092	0.0
4	Busseili #1	31°20′41″	30°25′55″	+ 0.0	2694	0.0
5	Mahmudiya #IX	31°08′42″	30°27′01″	+ 1.7	2405	0.0
6	Damanhur S. #IX	31°00′02″	30°27′23″	+ 3.2	2612	0.0
7	Dilingat, N. #IX	30°53′03″	30°29′14″	+ 3.3	2117	0.0
8	Rosetta #1	31°37′22″	30°32′34″	−22.0	3518	−22.0
9	Itay el-Barud #1	30°55′58″	30°38′53″	+ 1.0	2355	0.0
10	Abu Qir #IX	31°23′42″	30°13′47″	−14.0	3630	—
11	Sidi Salem #1	31°19′10″	30°43′16″	+ 2.2	4038	0.0
12	Baltim #1	31°41′57″	31°03′14″	−20.7	3910	—
13	Kafr el-Sheikh #1	31°10′23″	31°04′55″	+ 3.0	4182	0.0
14	Shebin el-Kom #1	30°33′26″	30°56′48″	+10.6	3214	0.0
15	Mit Ghamr #1	30°41′44″	31°16′26″	+ 9.5	2201	0.0
16	Abu Madi #1	31°25′45″	31°21′14″	+ 2.2	4163	0.0
17	Bilqas #1	31°10′12″	31°30′22″	+ 2.0	4571	0.0
18	el-Wastani #1	31°24′08″	31°35′46″	+ 0.4	3668	0.0
19	Abu Hammad #1	30°34′07″	31°50′04″	+40.5	4309	—
20	Ras el-Barr #1	31°40′27″	31°51′02″	−28.0	4444	—
21	Qawasim #1	31°20′07″	30°50′55″	− 3.0	3778	—
22	San el-Hagar #1	30°59′13″	31°50′53″	+ 1.6	3772	—
23	Qantara #1	31°01′36″	32°13′32″	0.0	4193	—
24	Abadiya #1	31°22′28″	31°03′10″	+ 1.0	3722	0.0
25	Matariya #1	31°12′22″	31°57′16″	+ 0.3	4142	0.0

[a] Q_3, Q_2, Quaternary Neonile and Prenile, respectively; Tplu, late Pliocene; Tpll, early Pilocene; Tm, Miocene; Tmu, late Miocene; To, Oligocene; Te, Eocene; K, Cretaceous; J, Jurassic. All boundaries are in meters above or below sea level.

Q₂	Tplu	Tpll	Tmu	Tm	To	Te	K	J
0.0	—	− 643	−1936	−2059	—	—	—	—
−21.8	− 195	− 308	−1532	−1618	−2156	—	−2325	—
−12.5	− 272	− 316	−1069	−1130	−2010	—	−2070	—
− 3.0	− 958	−1440	−2200	−2511	—	—	—	—
− 3.0	− 667	−1229	−1725	−1975	—	—	—	—
− 5.0	− 372	− 629	−1222	−1292	−2406	—	−2568	—
− 5.0	− 309	− 495	−1303	−1357	−1968	—	−2078	—
—	− 923	−1560	−2678	−2832	—	—	—	—
−13.0	− 370	− 829	−1519	−1660	−2261	—	−2319	—
0.0	− 832	−1076	−2464	−2545	—	—	—	—
−15.0	− 877	−2000	−2410	−3685	—	—	—	—
0.0	− 978	−2112	−3280	−3759	—	—	—	—
−14.0	− 975	−1880	−2735	−4048	—	—	—	—
−15.0	− 358	− 635	—	—	− 942	−1274	−1747	−3161
−20.0	− 483	− 663	—	—	−1014	−1701	−2001	—
−30.0	− 966	−2214	−3025	−3314	—	—	—	—
−25.0	−1000	−1986	−2751	−4087	—	—	—	—
−10.0	−1009	−2200	−2752	−2875	—	—	—	—
0.0	—	—	—	− 123	− 350	—	− 694	−1962
0.0	−1041	−2115	−2722	−2982	—	—	—	—
0.0	− 807	−1760	−2651	−3765	—	—	—	—
0.0	− 810	−1100	−1876	−2389	−3012	—	—	—
0.0	− 735	−1450	−1730	−2150	−3110	—	—	—
−22.0	−1169	−1723	−2787	−3295	—	—	—	—
−60.0	−1000	−1651?	−2290	−3330	—	—	—	—

Table A-2. Thickness and Elevation of Units Encountered in Selected Shallow Boreholes Drilled in the Nile Valley, Upper Egypt[a]

No.	Profile No. (W-E)	Well No.	Location	Elev. (m)	Total depth (m)
1	2	3	4	5	6
1	(4-4') Helwan sheet	21	Bernesht	+23	50
		22	Ezbet Hassanein	+23	300
		23	Ghammasa el-Kubra	+22	35
2	(5-5') Beni Suef sheet	24	Nazlet el-Saadna	+26	216
		25	Ezbet Marzouq	+26	71
		26	el-Shenawia	+28	33
3	(6-6') Minia sheet	27	Dakub	+35	355
		28	Istal	+35	495
		29	Nazlet Nakhla	+35	65
		30	Ebwan	+35	65
		31	Nazlet Abu Shehata	+37	349
4	(7-7') Malawi sheet	32	el-Arin	+42	154
		33	el-Roda	+43	106
		34	el-Beyadin	+43	39
5	(8-8') Tahta sheet	35	Um Doma	+56	56
		36	Nazlet Amer	+56	444
		37	Karam el-Arab	+56	396
		38	Hod Kaw	+57	397
6	(9-9') Naga Hammadi sheet	39	el-Karnak	+65	85
		40	Sidi Abdalla	+66	86
		41	Naga el-Kasab	+67	87
7	(10-10') Luxor sheet	42	Naga el-Berka	+75	107
		43	Naga Darau	+74	110
		44	el-Okb	+76	56
8	(11-11') Esna sheet	45		+82	24
		46		+80	92
		47	Hod el-Negout	+79	91
9	(12-12') Aswan sheet	48		+100	32
		49	Kom Ombo	+100	340
		50	Ezbet el-Yousefiya	+ 96	32
		51	Ezbet el-Atmour	+100	32

[a]Profiles are given in Figure 1.

Neonile (Q₃) top/thickness	Prenile (Q₂) top/thickness	Paleonile (Tplu) top/thickness	Eocene (Te) top/thickness	Nubia sandstone (Kn) top/thickness
7	8	9	10	11
+23/20 +23/15 +22/10	+ 3/30 + 8/25 +12/25	−17/130	−147/130+	
+26/10 +26/10 +28/10	+16/206 +16/61 +18/23			
+35/15 +35/10 +35/10 +35/10 +37/10	+20/244 +25/285 +25/16 +25/16 +27/32	−224/96 −260/200 — — − 5/317	+ 9/39+ −30/39+	
+42/12 +43/12 +43/20	+30/142 +31/94 +23/19			
+56/ 4 +56/ 4 +56/10 +57/ 8	+52/52 +52/88 +46/126 +49/129	−36/352 −80/260 −80/260		
+65/20 +66/ 8 +67/16	+45/65 +58/78 +51/71			
+75/28 +74/20 +76/12	+47/79 +54/90 +64/44			
+82/ 8 — +79/ 8	+74/16 +80/92 +71/83			
+100/4 +100/4 + 96/4 +100/4	+96/28 +96/28 +92/28 +96/28	+68/224		−156/84

**Table A-3. Thickness and Elevation of Measured Sections of
Paleonile (Tplu), Armant, and Issawia Formations (Tplu/Q₁
Interval)**[a]

No.	Location	Thickness (m)			Elevation of top (m)
		Tplu	Armant	Issawia	
1	2	3	4	5	6
1	Wadi Um Sulimat	—	39	—	180
2	Wadi Um Sulimat	—	10.8	—	180
3	Wadi Um Sulimat	—	33.0	—	210
4	Wadi Um Sulimat	—	13.5	—	210
5	Wadi Um Sulimat	—	12.0	—	210
6	Higab el-Rakham	—	17.0	—	230
7	Higab el-Rakham	—	11.5	—	210
8	Wadi el-Surai	—	30.5	—	137
9	Wadi el-Surai	—	16	—	154
10	Wadi el-Surai	28.5	5	—	150
11	Wadi el-Surai	—	15	—	150
12	Wadi el-Surai	15	6	—	180
13	Wadi el-Surai	22.5	—	—	180
14	South Gebel el-Gir	9	—	—	180
15	Wadi el-Suweini	23	4	—	170
16	Wadi Abu Sakrana	—	8.5	—	122
17	Wadi el-Matuli	8.5	—	—	120
18	Wadi el-Matuli	8.5	—	—	120
19	Wadi el-Matuli	11.1	—	—	150
20	Wadi el-Matuli	4.75	—	—	120
21	Wadi el-Qreiya	10	18	—	210
22	Wadi el-Matuli	15	6	—	150
23	Wadi el-Qreiya	10.5	—	—	150
24	Wadi Aras	20.5	12	—	230
25	Wadi Aras	6.2	8.1	—	190
26	Wadi Aras	28.0	0.5	—	190
27	Wadi Aras	25	6.0	—	200
28	Wadi Aras	13.0	—	—	150
29	Wadi el-Tameid	12.0	1.0	—	160
30	West Gebel Aras	26.6	18	—	120
31	West GebelAras	11	12	—	170

Table A-3. (*Continued*)

No.	Location	Tplu	Armant	Issawia	Elevation of top (m)
		Thickness (m)			
1	2	3	4	5	6
32	Wadi el-Sheikh Omar	10	—	—	150
33	Wadi el-Sheikh Ali	13.2	16	—	150
34	Wadi Abu Nafukh	5.4	4	—	200
35	Wadi Abu Nafukh	4	17.5	—	200
36	Wadi Abu Nafukh	—	24	—	200
37	South Wadi Matahit	20	—	—	150
38	Wadi Aras	20	5	—	—
39	Wadi Aras	10.6	11	—	—
40	Wadi Aras	31.25	6.6	—	—
41	West Gebel Abu Had	5.7	8.7	—	—
42	Wadi Abu Had	9.0	3.7	—	—
43	El Heita	19	12	—	—
44	Wadi Aras	—	32.4	—	—
45	Wadi Aras	—	22.4	—	—
46	Wadi Aras	—	36.5	—	—
47	Wadi Gurdi	—	12.4	—	—
48	Wadi Gurdi	10.5	8.0	—	—
49	Wadi el-Nuzeiza	4.5	—	14.4	140
50	Wadi el-Kiman	12	—	34.2	131
51	Issawia quarry	15.75	—	29.25	131
52	Wadi Abu Azzuz	9.0	—	3.0	120
53	Wadi Banat Birri	27.3	4.5	—	166
54	Wadi el-Madamud	2.65	26.75	—	180
55	Wadi el-Madamud	37.45	—	—	180
56	Wadi el-Madamud	30.75	1.1	—	176
57	Wadi el-Madamud	—	13.15	—	166
58	South Luxor	23.5	2	—	150
59	North Gebel el-Gir	—	29.95	—	180
60	Wadi Abu Suba	—	16	—	153
61	Wadi Arqub el-Baghla	—	39.7	—	180
62	Wadi el-Bairiya	—	39.2	—	180
63	Wadi el-Direira	11.3	1	—	180

No.	Location	Thickness (m)			Elevation of top (m)
		Tplu	Armant	Issawia	
1	2	3	4	5	6
64	Wadi el-Bairiya	—	39.7	—	210
65	South el-Qusiya	—	—	11.2	102
66	East Beni Adi	11.9	—	9.15	110
67	South Sohag	12.1	10.5	—	147
68	East Gebel Abu el-Nut	4.2	—	—	127
69	Wadi Samnud	11	5.1	—	142
70	Wadi Beni Himeil	7.0	8.5	—	142
71	North Kolet el-Qataya	14.0	—	4.5	143
72	South Kolet el-Qataya	10.0	1	—	146
73	West el-Qata	2.0	1.5	—	102
74	South Naqb el-Hoggag	10.8	6.25	—	119
75	West Naqh el-Hoggag	10.3	5.0	—	90
76	Wadi Abu Shih	14.3	3.8	—	99
77	North Wadi Abu Shih	6.0	—	5.0	120
78	North Wadi Abu Shih	7.1	—	3.0	120
79	North Noqb el-Adabiya	—	22.1	—	143
80	Wadi Abu Shih	20.7	—	—	90
81	South West Gebel Tarbul	—	3.75	—	62
82	Wadi Mitin	—	6.6	—	87
83	South West Gebel Tarbul	—	3.2	—	62
84	Gebel Um Raqaba	19.2		marine Pliocene	123
85	Wadi Sannur	—	1.9	—	146
86	Wadi Sannur	—	10.9	—	149
87	Wadi el-Assiuti	—	8.05	—	92
88	Wadi el-Assiuti	—	6.0	—	90
89	Wadi el-Assiuti	—	15.8	—	116
90	Wadi el-Assiuti	—	3	—	150
91	Wadi el-Assiuti	7.4	2.9	—	90
92	Wadi el-Assiuti	13.4	8.6	—	108
93	Wadi el-Assiuti	13.2	9.6	—	96
94	Wadi el-Assiuti	17.35	—	3.5	97
95	East Mallawi	16.0	0.5	—	75

[a]See Figure 18 for logs of sections and Figure 19 for their locations.

**Table A-4. Thickness and Elevation of Measured Sections of
Prenile (Q₂) Sediments**

No.	Location	Thickness (m)		Elevation of top (m)
		Qena	Abbassia	
1	2	3	4	5
101	Wadi el-Surai	8.0	4.0	120
102	Wadi el-Surai	10.5	2.2	115
103	Wadi el-Surai	14.0	4.5	120
104	Wadi el-Surai	10.0	3.5	120
105	Wadi el-Surai	15.8	4.0	120
106	Wadi el-Surai	16.2	6.0	160
107	Wadi el-Surai	11.0	1.0	120
108	Northwest Al-Ashraf	10.0	4.0	100
109	Northeast el-Gabalaw	11.0	6.5	110
110	South Gebel el-Gir	17.7	4.0	180
111	North east el-Gabalaw	4.5	0.3	80
112	Wadi el Shuweini	17.5	5.0	150
113	Wadi el Shuweini	13.0	—	130
114	West Gebel el-Gir	15.0	4.0	120
115	West Gebel el-Gir	3.0	2.0	160
116	East Wadi Qena	18.0	—	120
117	East Wadi Qean	—	8.0	85
118	Wadi el-Sheikh Eda	13.0	1.5	131
119	Wadi el-Sheikh Masallat	11.1	2.3	116
120	West Wadi el-Masallat	6.5	2.0	120
121	Wadi el Masallat	2.8	1.5	103
122	Wadi el-Masallat	7.5	1.5	120
123	Wadi el-Matuli	—	8.5	116
124	Wadi Aras	9.0	12.0	170
125	Wadi Aras	11.0	4.0	150
126	Wadi Aras	14.0	7.5	150
127	Wadi el-Higeirat	—	14.0	120
128	West Wadi Qena	19.5	2.5	131
129	Wadi el-Shahadein	24.0	9.0	150
130	North el-Makhadma	12.5	4.0	117
131	Wadi el-Qenawiya	21.4	—	128

Table A-4. (*Continued*)

No.	Location	Thickness (m)		Elevation of top (m)
		Qena	Abbassia	
1	2	3	4	5
132	Wadi el-Sheikh Omar	25.0	4.0	144
133	Wadi el-Zineiqa	7.5	5.0	111
134	Wadi el-Sheikh Ali	21.0	4.5	115
135	North el-Hisha	1.0	1.5	98
136	North el-Hisha	8.0	3.0	98
137	Wadi el-Khaiyata	20.5	3.5	82
138	Wadi el-Khaiyata	9.0	4.0	82
139	Wadi el-Khaiyata	2.0	0.5	82
140	Wadi Qasab	19.2	—	90
141	Wadi Qasab	5.0	6.0	101
142	Wadi Abu Nafukh	24.7	—	124
143	Wadi Matahir	1.5	1.5	85
144	South Wadi Matahir	16.0	1.0	140
145	South Wadi Matahir	18.0	1.0	106
146	South Wadi Matahir	8.2	2.5	92
147	El Maana quarry	10.5	—	109
148	Wadi el-Zineiqa	12.5	12.5	140
149	Wadi el-Kiman	9.0	1.0	102
150	Wadi Saflak	19.3	2.2	—
151	Wadi Abu Azzuz	4.5	4.0	—
152	South Wadi Banat Beirri	9.2	5.0	122
153	South Wadi Abu Ashar	25.6	12.0	112
154	East Gebel el-Gir	5.5	3.0	122
155	Wadi Imran	5.5	3.0	123
156	North Gebel el-Gir	17.9	13.5	122
157	Wadi el-Rimeila	9.1	7.7	100
158	Wadi el-Heba	15.6	12.0	120
159	Wadi el-Heba	13.6	—	138
160	Wadi el-Bairiya	19.6	3.5	110
161	Wadi el-Rimeila	7.4	3.9	100
162	West el-Minia	2.0	1.7	61

Table A-4. (*Continued*)

No.	Location	Thickness (m) Qena	Abbassia	Elevation of top (m)
1	2	3	4	5
163	West el-Minia	3.7	0.5	61
164	West Samalut	2.7	—	49
165	West Samalut	3.6	—	49
166	West el-Minia	0.5	0.25	64
167	South west el-Minia	—	0.6	64
168	West Samalut	1.3	0.25	49
169	Beni Adi	8.6	3.7	49
170	South West el-Kawami	7.7	2.0	92
171	West Sohag	3.5	0.5	111
172	Wadi Samhud	—	9.0	102
173	Wadi Samhud	1.8	4.5	102
174	North Wadi Samhud	9.0	—	108
175	Wadi Beni Himeil	10.0	—	95
176	North Wadi Abu Shih	2.3	0.5	84
177	Wadi Abu Shih	4.2	—	86
178	South Wadi Abu Shih	5.0	13.0	86
179	South Wadi Abu Shih	6.3	4.5	86
180	Wadi el-Assiuti	4.4	0.8	92
181	Wadi el-Assiuti	5.0	1.7	95
182	Wadi el-Assiuti	12.5	0.4	95
183	Wadi el-Assiuti	23.6	1.85	95

[a]See Figure 19 for locations of sections.

ABBASSIA FORMATION
Middle Pleistocene (Q$_2$)

Type section
Quarry of gravel pit at Rus Station halfway along the Faiyum-Wasta railway line.

Lithology
Massive loosely consolidated gravels of polygenetic origin. The pebbles are rounded to subrounded and were transported from the uncovered basement of the Eastern Desert of Egypt.

Stratigraphic limits
Overlies unconformably the Qena sands. The upper limit has not been observed.

Extent
The formation is widely spread, along the Nile Valley and edges of the delta. In Nubia it forms inverted ridges in the Nubian Desert.

Fossils and other datable materials
Includes abundant late Acheulian hand axes. The formation represents the deposits of an ephemeral river which succeeded the Prenile and had its sources in the Eastern Desert of Egypt. The age is middle Pleistocene (Q$_2$).

Reference
Said (1975).

ABU MADI FORMATION
Early Pliocene (Tpll)

Type section
Abu Madi well # 1 from depth 2007 to 3129 m. The coordinates of the well: lat. 31°26'17" N; long. 31°21'14" E.

Lithology
The formation is made up of thick layers of sands (rarely conglomeratic) interbedded with clay beds which become thicker and more frequent in the lower part of the formation. The sand is quartzose, poorly sorted, and loose to poorly cemented. The conglomeratic layers are more frequent in the lower part of the formation and have a sandy matrix.

Stratigraphic limits
Overlies the Rosetta evaporites and underlies the Kafr el-Sheikh formation which is made up almost exclusively of clay beds. In some places, however, the upper part of the Abu Madi includes clay beds while the lower part of the Kafr el-Sheikh includes more sand beds and in this case the differentiation of the upper limit of the Abu Madi becomes difficult.

Extent
North Delta embayment in the subsurface.

Fossils and age
Foraminifera are fairly frequent in the clay interbeds. They belong to the *Sphaeroidinellopsis* zone. In places the lower part of the formation is non-fossiliferous massive and could be of subaerial origin. The upper part is definitely marine in origin and is of early Pliocene age.

Reference
Rizzini *et al.* (1978)

ARKIN FORMATION
Holocene

Type section
Site D-1 of the Combined Prehistoric Expedition, Arkin, Wadi Halfa district, Sudanese Nubia.

Lithology
The formation is made up of silts and fine-grained micaceous sand (= younger Neonile deposits).

Extent
Known in outcrop only in Nubia where it is represented by embankments of silt about 13 m higher than the modern Nile level (prior to the erection of the Aswan High Dam). In Egypt the younger Neonile deposits are buried under the alluvial plain.

Fossils and age
Radiocarbon dates in Nubia date the lower part of the Arkin between 10,000 and 7750 B.P. There it carries fossil mollusks (*Unio willcocksi* and *Corbicula vera*) which disappear in the upper part of the section which is separated by de Heinzelin (1968) into a unit which he terms the Qadrus.

Remarks
As originally defined by de Heinzelin in 1968 in Nubia, the Arkin covers the earlier part of the deposits of the Holocene while the Qadrus covers the later part. Both units have not been observed in outcrop in Egypt where the Holocene deposits are collectively and loosely classified under the Arkin.

ARMANT FORMATION
Early Pleistocene (Q₁)

Type section
Wadi Bairiya, opposite Armant, Luxor district.

Lithology
The formation is made up of alternating beds of locally derived gravels and fine-grained clastic rock. The gravel beds are cemented by tufaceous material and the pebbles are subangular and poorly sorted. The fine-grained clastic beds are calcareous, sandy, shaly, or phosphatic depending on the nature of the nearby source rock.

Stratigraphic limits
Overlies older bedrock and underlies the Qena formation. The formation abuts against the sides of wadis draining into the Nile or forms inverted wadi ridges.

Extent
The formation fills many of the wadis that drain into the Nile in Upper Egypt.

Fossils and age
The formation is nonfossiliferous and is considered on stratigraphic relations as of early Pleistocene age representing the deposits of an early pluvial, the Armant.

Reference
Said (1975).

BALLANA FORMATION
Late Pleistocene (Q₃)
Formation proposed by de Heinzelin (1968) for the windblown sands interfingering the top parts of the β aggradational silts. See Masmas-Ballana Formation.

BASAL NEONILE DEPOSITS
See Neonile Deposits

BILQAS FORMATION
Holocene

Type section
Bilqas well #1 between depths 0 and 25 m. The coordinates of the well: lat. 31°10'12" N; long. 31°30'22" E.

Lithology
The formation is made up of alternating fine to medium-grained sand and clay interbeds. The clay beds include vegetable matter and peat beds.

Extent
North delta regions in the subsurface, probably forming in swamp or coastal environment.

Fossils and age
Recent Mediterranean mollusks, Holocene.

BIRBET FORMATION
Late Pleistocene (Q₃)
Name proposed by de Heinzelin (1968) for deposits built up during the recessional episode separating the Sahaba and Arkin aggradations (around 11,000 B.P.) in Nubia. Formation made up of slope-wash debris underneath the younger Neonile deposits.

DANDARA FORMATION
Middle Pleistocene (Q₂)

Type section
Dandara, 3 km south of the Temple of Hathor, west bank, Qena province.

Lithology
The formation is made up of a lower gray calcareous sandy silt bed and an upper unit of brown silt with a

number of thin limestone interbeds and occasional lenses of locally derived gravel.

Stratigraphic limits
Overlies the Qena sands and underlies the Ikhtiariya dunes.

Extent
Occurs as isolated discontinuous patches along the Nile banks in Upper Egypt and as a more continuous embankment in Nubia.

Fossils and datable materials
Radiocarbon date of the upper limestone bed gives an age older than 40,000 B.P. Upper levels carry one implement with an evolved Levalois technique. The formation represents the deposits of the earliest Neonile (α Neonile) with a regimen similar to the present river. The age is upper Pleistocene.

Reference
Said *et al.* (1970).

DEIR EL-FAKHURI
Late Pleistocene (Q_3)

Type section
Site E71 K14 of the Combined Prehistoric Expedition, west bank, Esna, Upper Egypt.

Lithology
Pond sediments and diatomites with interfingering thin silt laminae. Thickness 2 m. The diatomites are sandy and highly calcareous.

Stratigraphic limits
Overlies the Ballana dunes and underlies the γ aggradational silts (Sahaba-Darau formation).

Extent
Recorded only in type locality and in Tushka area in Egyptian Nubia. Other recessional features associated with this recessional episode are the vegetation casts observed at the top of the Ballana dunes and the noncalcic soil which developed over these dunes.

Fossils and other datable materials
The diatoms described from this unit are cosmopolitan in aspect and they include three species which may have tropical affinities (*Navicula cuspidata, Gomphonema lanceolatum,* and *Ephitemia sorex*). This contrasts with the Holocene diatom flora of the Faiyum, where tropical species are common. Some of the species recorded prefer deep water (*Coconeis placenta* and *Melosira granulata*). The pollen and spores are mostly boreal species, but may suggest grassland rather than aquatic environment. A sample of carbonaceous sand at the base of the diatomite yields a radiocarbon date of 12,700 B.P.

References
Said *et al.* (1970); Wendorf and Schild (1976); Albritton (1968).

DIBEIRA-JER FORMATION
Late Pleistocene (Q_3)
As originally described by de Heinzelin (1968) from Sudanese Nubia, the Dibeira-Jer includes the lower silts and fluviatile sands of the sediments of the Nile with modern aspect. They were correlated with the lower silts of the Esna section, Upper Egypt, which carry late Paleolithic artifacts. The Dibeira-Jer was restudied by Wendorf and Schild (1976) and found to be formed of at least two aggradational units, a lower carrying middle Paleolithic artifacts and an upper which underlies the γ Aggradational silts (Sahaba formation). In the present work the lower silts isolated by Wendorf and Schild are correlated with the α Neonile deposits (Dandara) and the upper silts with the β Aggradational silts (Masmas-Ballana).

DISHNA FORMATION
See Dishna-Ineiba Formation.

DISHNA-INEIBA FORMATION
Late Pleistocene (Q_3)

Type section
Dishna member: Dishna, east bank, Qena province. Ineiba member: new Ineiba village, Kom Ombo plain. Name used in this book for the playa and wadi deposits which formed during the recession following the Sahaba-Darau aggradation and also during the final Paleolithic-Neolithic pluvial.

Lithology
The Dishna member is coeval with the Malki member of Butzer and Hansen (1968) and is made up of yellowish brown clays which follow on top or interfinger the last phases of the Sahaba-Darau aggradation. Dates of this event are between 12,000 and 11,000 B.P.

The Ineiba member is a playa deposit made up of brown clays and interfingering gravel beds of local derivation. These probably formed during the earlier phases of the younger Neonile δ Aggradational silts continuing to the final Paleolithic-Neolithic pluvial (9000–7000 B.P.).

References
Butzer and Hansen (1968); Said *et al.* (1970).

GAR EL-MULUK FORMATION

Type section
Gar el-Muluk hill, Wadi Natrun depression.

Lithology
A 35-m section of sand, shale, and a few limestone bands.

Stratigraphic limits
Ostrea cucullata in lower levels, fluviomarine vertebrates and ostracods in upper levels. The age is late Pliocene representing the deltaic deposits of the Paleonile.

References
Blanckenhorn (1901); Said (1962a); James and Slaughter (1974).

GEBEL SILSILA FORMATION
Late Pleistocene (Q₃)

Name proposed by Butzer and Hansen (1968) for the Aggradational silts of the Neonile. The type section is in the northeastern reaches of the Kom Ombo plain where an old channel of the Neonile filled the Silsila gap. The lithology of these channel beds is sand with pebble beds whereas the lithology of the levees is sandy silts with root drip and carbonized vegetable matter. The formation is rich in *Unio* and *Cleopatra* shells. It is dated between 12,000 and 10,000 B.P. and is coeval with the Sahaba-Darau formation.

GIRRA FORMATION
Holocene

Type section

Lithology and stratigraphic limits
Nilotic silts occurring as embankments of great extent about 5 m higher than the flood plain level of the Nile at Debba, Sudanese Nubia.

Fossils and other datable materials
Accompanied by lateral pediments (10 YR) on which rest various deflated early ceramic sites. It is thus assumed to be before 7000 years B.P. which suggests an equivalence to the Arkin formation north of the 2nd cataract.

Extent
Sudanese Nubia, Dongola reach.

Reference
Shiner *et al.* (1971).

GOSHABI FORMATION
Late Pleistocene

Type section
Site N2-1 of the Combined Prehistoric Expedition, Debba, Sudanese Nubia.

Lithology and stratigraphic limits
Nilotic deposits made up of gravel overlying fluviatile sand which rests in turn on hardened silt with concretions. They form embankments about 13 m above the level of the present flood plain in the Debba region, Sudanese Nubia.

Fossils and other datable materials
While mostly sterile, artifacts were found *in situ* in both the fluviatile sand and at the contact between the fluviatile sand and silt. The age is late Pleistocene. They are correlated on lithological and archeological evidence with the Neonile deposits.

Extent
The Nile at Debba at the mouth of Wadi el-Milk, Sudanese Nubia.

Reference
Shiner *et al* (1971).

HELWAN FORMATION
Late Pliocene (Tplu)

Type section
Wadi Garawi, south of Helwan, south Cairo.

Lithology
The formation is made up of a long series of rhythmically banded alluvial silts, fine-grained sands, and marl interbeds. The thickness of the formation exceeds 120 m.

Stratigraphic limits
Overlies upper Eocene bedrock and underlies the high gravel terraces of the Idfu formation.

Extent
Occurs along the sides of the valley and many of the wadis which drain into the Nile.

Fossils and age
The formation carries a limited number of species of mollusks, the most common of which is *Melanopsis* sp. The age is late Pliocene, correlated with the Paleonile fluvial sediments.

References
Blanckenhorn (1921); Said (1962a, 1971).

IDFU FORMATION
Early Pleistocene (Q₁)

Type section
Pit #1, Wadi el-Hassayia, Darb el-Gallaba plain, Idfu west, Upper Egypt.

Lithology
The formation is made up of gravels and sands of fluvial origin embedded in a red-brown matrix. The sands are quartzose and the gravels are mostly flint. The thickness is unknown but probably exceeding 20 m.

Extent
The formation is known along the western banks of the Nile and delta at elevations reaching 90 to 100 m above the modern flood plain of the river.

Stratigraphic limits
The stratigraphic position of this formation is *not* known with certainty, but it is assumed to overlie the Paleonile sediments and to be of early Pleistocne age (Q₁). The Idfu is definitely older than the Prenile formations. It is nonfossiliferous and archeologically sterile. It represents the deposits of the Protonile.

Reference
Said (1975).

IKHTIARIYA FORMATION
Early Pleistocene (Q₁)

Type section
Site 34 of the Combined Prehistoric Expedition, Dibeira west, Sudanese Nubia.

Lithology
Dune sand, massive and well-sorted. Thickness is in the range of 4 to 6 m.

Stratigraphic limits
Overlies the eroded bedrock or the Korosko Formation with which it also interfingers, and underlies unconformably the Masmas-Ballana.

Extent
Nubia and Upper Egypt; the top of the formation is 13 m above the flood plain level in Nubia (before the erection of the Aswan High Dam) and is barely at the same level of the modern flood plain of Upper Egypt.

Fossils and other datable materials
Middle Paleolithic implements, few mammal bones.

Age
Assumed to represent the eolian deposits contemporaneous with and succeeding the Mousterian-Aterian pluvial (80,000–40,000 B.P.).

Reference
de Heinzelin (1968).

INEIBA FORMATION
Late Pleistocene
Name proposed by Butzer and Hansen (1968) for the alluvial wadi deposits formed during and after the Gebel Silsila formation. It is divided into the lower Malki and the upper Sinqari members which may be contemporaneous with the Dishna and Ineiba members of the Dishna-Ineiba formation. See Dishna-Ineiba Formation.

ISSAWIA FORMATION
Early Pleistocene (Q₁)

Type section
Issawia quarry face, Akhmim, Upper Egypt.

Lithology
The formation is made up of bedded travertines with minor conglomerate lenses with pebbles of local derivation and an overlying bed of red breccia with highly angular limestone pebbles cemented in a matrix of red muds.

Extent
The Nile Valley in its middle latitudes at localized areas.

Stratigraphic limits
Overlies the Armant formation.

Fossils and age
Unidentified reeds and stems. The formation is considered lower Pleistocene.

Reference
Said (1975).

KAFR EL-SHEIKH FORMATION
Late Pleistocene (Tplu)

Type section
Kafr el-Sheikh well #1 from depth 975 to 2735 m. The coordinates of the well: lat. 31°10'23" N; long. 31°4'55" E.

Lithology

The formation is made up of clay beds with minor sand interbeds. The sand is quartzose held together by a clay matrix. The clays are made up of equal proportions of kaolinite and montmorillonite with very little illite.

Extent

In the subsurface in all delta and valley wells with remarkable constancy.

Stratigraphic limits

Overlies the Abu Madi and underlies with a marked unconformity the Wastani sands in the North Delta embayment or the graded sand–gravel beds of the Prenile (Q_2) in the southern reaches of the delta and the valley.

Fossils and age

In the northern wells the formation carries a rich marine fauna of the early and late Pliocene. In the southern wells of the delta and the valley, the marine faunas give way to brackish water forms indicating the dominance of fluvial over marine conditions. The age of the fluvial sediments related to the Paleonile is late Pliocene. In the type section the upper 905 m (between depths 975 and 1880 m) belong to the late Pliocene.

Reference

Rizzini *et al* (1978).

EL-KHEFOUG FORMATION
Late Pleistocene-Holocene

Type section

el-Khefoug, north of Ihnasia el-Medina, west bank, Beni Suef province.

Lithology

Made up of alternating nilotic silts and wind driven sands.

Stratigraphic limits

Overlies the Prenile sediments.

Extent

Occurs all along the west bank of the Nile from Mallawi to Ihnasia el-Medina for a distance of 150 km.

Fossils and age

Nonfossiliferous, but carries archeological materials of late Paleolithic to Neolithic ages. The formation is made up mainly of interbedded dune sand and silts of the Sahaba γ Neonile aggradation. The upper parts

of the formation seem to belong to more recent dune laminae interbedded with silts of the Holocene δ Neonile deposited in exceptionally high floods. The rising floods stabilized the dunes and prevented them from overwhelming the cultivation. The age is late Pleistocene-Holocene.

KOM EL-SHELUL FORMATION
Early Pliocene (Tpll)

Type section

Kom el-Shelul, Pyramids plateau, Gizeh.

Lithology

The formation is made up of sandstone and coquinal limestone beds.

Stratigraphic limits

The formation overlies unconformably the upper Eocene Maadi formation and underlies unconformably the Idfu gravels.

Extent

Abuts with a depositional dip against the sides of the excavated Nile Valley from Cairo to Beni Suef.

Fossils and age

Chlamys scabrella, *Strombus coronaturs*, *Clypeaster aegyptiacus*, and a rich foraminiferal and ostracod fauna.

References

Sandford and Arkell (1939); Said (1962a).

KOROSKO FORMATION
Late Pleistocene (Q3)

Type section

Wadi Ayed, New Korosko village, Kom Ombo plain.

Lithology

Described by Butzer and Hansen (1968) as a sandy–marly unit forming what these authors termed the "Basal Sands and Marls" of the Kom Ombo nilotic succession. The unit was interpreted as of nilotic (p. 87), lacustrine or mixed fluvial-semiaquatic origin (p. 95), or due to wadi activity (p. 91). A restudy of this unit showed that it underlies the older Neonile deposits and that it was formed as a result of wadi activity by sheet floods. The unit is made up of a lower bed of subangular pebbles of local derivation embedded in a matrix of coarse sand and an upper bed of light gray to light brownish gray (2.5 Y 6-7/2) poorly sorted medium sand to sandy marl with about 15–40% $CaCo_3$. The unit owes its origin to deposition by torrential rains at the mouth of wadis which drained the Eastern Desert of Egypt.

Stratigraphic limits
Underlies unconformably the Ikhtiariya dunes or the Masmas-Ballana silts and overlies bedrock or older Nile deposits.

Extent
Known mostly in pits along the east bank of the river in Upper Egypt. Few expsoures have been noted in Sebaiya, Idfu, and Kom Ombo. Wendorf and Schild (1976) described many of these occurrences in their work on the east bank of the Nile in Upper Egypt (e.g., trenches at sites E61 M 2–10, Dishna, Qena province). Along the west bank Sandford and Arkell's (1929) 10-foot terrace belongs to the Korosko formation. A critical section is between Medinet Habu and the Coptic Church, Luxor where gravels of local derivation form a surface of great extent.

Fossils and other datable materials
Mousterian implements were recorded from the 10-foot terrace of Sandford and Arkell from near Luxor. Butzer and Hansen separated several molluscan species from the alluvial deposits of the formation (*Bulinus, Planorbis, Valvata*, etc.). The formation represents the deposits of the long recessional episode which separated the α Neonile from the succeeding β Neonile.

MADAMUD FORMATION
Late Pliocene (Tplu)

Type section
Wadi Madamud, east bank, Qena province.

Lithology
Chocolate brown marls and rhythmically banded fine sand and silt laminae.

Stratigraphic limits
Overlies older bedrock and underlies the Armant formation.

Extent
Abutting against the sides of many wadis draining into the Nile from the Eastern Desert of Egypt.

Fossils and age
Nonfossiliferous but related to the fluvial deposits of the Paleonile and coeval with the Helwan formation. The age is late Pliocene.

MAKHADMA FORMATION
Late Pleistocene (Q₃)

Type section
Makhadma, east of the Nile opposite Dandara, Qena province.

Lithology and stratigraphic limits
Slope wash forming over the rolled topography of the Prenile and Abbassia sediments.

Extent
Upper Egypt along the east bank of the river.

Fossils and other datable materials
Carries archeological materials of middle Paleolithic tradition, probably contemporaneous with the Korosko formation. It was formed during the wet Mousterian-Aterian interval occurring during the recession separating the α Neonile from the β Neonile.

References
Said (1975); Wendorf and Schild (1976).

MALKI MEMBER
See Ineiba Formation.

MASMAS FORMATION
See Masmas-Ballana Formation.

MASMAS-BALLANA FORMATION
Late Pleistocene (Q₃)

Type section
Site E71 K1 of the Combined Prehistoric Expedition, Esna west bank.

Lithology
The formation is made up of nilotic silts (Masmas) interfingering with dune sands (Ballana). The lower part is usually made up of a solid unstratified sandy clayey silt with medium angular blocky structure and rare slicken sides. The upper part is made up of yellowish medium to coarse-grained, loose, calcareous dune sand with forest beds dipping 20–40% to the east. Rare oxidation stains are concentrated mainly near the top. Very numerous root drip casts occur near the top from the dune surface. Human occupation remains in the form of chipped stone artifacts; animal and fish bone litter the surface and are also dispersed within the top 20 cm of the dune. The dune has thin (0.5–2 cm) continuous interfingering streaks of sandy silt as well as thick lenses of nilotic silts. In the Esna region the base is unexposed and the thickness of the Masmas-Ballana does not exceed 3 m. In Kom Ombo, where the

Masmas overlies the Korosko, the formation attains a thickness of more than 25 m.

Stratigraphic limits
Overlies the Korosko or the Ikhtiariya formations and underlies the Deir el-Fakhuri or the Sahaba-Darau formations. The formation occurs on embankments lying about 8 m above the flood plain level.

Extent
Known along both banks of the Nile in Upper Egypt. Along the east bank, the dune element of the formation becomes less frequent and even wanting.

Fossils and other datable materials
Molluskan fauna collected from this information include *Planorbis ehrenbergi*, *Valvata nilotica*, *Bulinus truncatus*, and rare *Corbicula fluminalis*. No vertebrate remains have been found in this unit, but the archeological materials are of late Paleolithic tradition and are found in the top part of the Ballana dune.

Age
Forming the β Neonile aggradational deposits starting from about 30,000 and terminating around 18,000 B.P. (the latter date is based on a radiocarbon date of a *Unio* shell picked from the litter of human occupation above the dune).

Synonym
Lower silts of Said, Wendorf, and Schild (1970); Dibeira-Jer of de Heinzelin (1968).

Reference
Wendorf and Schild (1976).

MELANOPSISSTUFE
Late Pliocene (Tplu)

Name proposed by Blanckenhorn (1901) for the old Diluvium deposits now given the name Helwan formation.

MIT GHAMR FORMATION
Middle Pleistocene (Q₂)

Type section
Mit Ghamr well #1 from depth 20 to 483 m. The coordinates of the well: lat. 30°41'44" N; long. 31°16'26" E.

Lithology
The formation is made up of sand and pebble beds with a few minor clay interbeds in the lower part.

The pebbles are quartz, chert, or silicified limestone. The sands include molluscan shells forming coquina at certain levels. There are a few peat horizons.

Stratigraphic limits
Overlies the el-Wastani sands which could be part of this unit and underlies the agricultural clay silt layer of the land of Egypt.

Extent
The formation is recorded in all wells drilled in the delta. In the north it carries marine faunas, whereas in the south it is nonfossiliferous. The name is used for the subsurface marine section of the northern reaches of the delta. The formation is coeval with the Prenile Qena sands. The age is middle Pleistocene (Q_2).

NEONILE DEPOSITS

Deposits of the last and extant river occupying the valley of the Nile, made up of four units of aggradational deposits formed by four succeeding rivers, the α, β, γ, and δ Neoniles, separated by three recessions, the α/β, β/γ, and γ/δ. The lowest recessional episode, the α/β, was the longest and included the Mousterian-Aterian pluvial. The Neonile deposits are divided into three major units which are from top to bottom:
(1) The younger Neonile deposits including the deposits of the δ Neonile and the preceding recessional deposits (the γ/δ or the Dishna-Ineiba).
(2) The older Neonile deposits including the deposits of the γ Neonile (Sahaba-Darau), the β Neonile (Masmas-Ballana), and the intervening β/γ recession (Deir el-Fakhuri).
(3) The Basal Neonile Deposits including the deposits of the α Neonile (Dandara) and the α/β recession (made up of the deposits of the arid Gerza-Ikhtiariya and wet Korosko-Makhadma episodes).
The age of the deposits is late Pleistocene to Recent.

QADRUS FORMATION
See Arkin Formation

QAWASIM FORMATION
Late Miocene (Tmu)

Type section
Qawasim well #1 from depth 3765 to 2651 m. The coordinates of the well: lat. 31°20'07" N; long. 30°50'55" E.

Lithology

The formation is made up of thick layers of sands, sandstones, and conglomerates interbedded with thin clay layers. The sands are poorly sorted, loosely cemented by a clayey carbonaceous cement. The conglomerate layers are thick and include cobbles and boulders derived from the Cretaceous and Eucene landmass of Egypt.

Stratigraphic limits

Overlies the Sidi Salem formation and underlies the Rosetta evaporites.

Extent

North Delta embayment in the subsurface.

Fossils and age

The fauna is scarce and is found at certain localized horizons. When found the fauna is rich in number of individuals but poor in the number of species. The Ostracoda and Foraminifera indicate a late Miocene age and a fluvial to deltaic environment with minor marine incursions.

Reference

Rizzini *et al.* (1978).

QENA FORMATION
Middle Pleistocene (Q₂)

Type section

Wadi Abu Manaa quarry face, east bank, Dishna, Qena province.

Lithology

The formation is made up of cross-bedded fluvial sands with minor conglomerate and clay beds. At the type locality the base is unexposed and the thickness exceeds 20 m.

Stratigraphic limits

The lower limit of the formation is observed in outcrop in Wadi Abu Nafukh, east of Qena, Upper Egypt, where it overlies the late Pliocene Paleonile sediments (Madamud formation). The formation seems to be coeval with the graded sand–gravel bed found beneath the agricultural clay layer of the land of Egypt which overlies the deposits of the Paleonile (Kafr el-Sheikh formation). The composite thickness of the unit including the subsurface graded sand–gravel unit exceeds 70 m in Upper Egypt. The formation underlies unconformably the Abbassia gravels.

Extent

Nile Valley and delta. In the south forming part of the rolling topography of the Nile Valley; in the north forming embankments of great extent along the valley and the Faiyum depression; also known along the delta edges. The Qena sands represent the deposits of a competent river which preceded the modern Neonile.

Fossils and age

Corbicula shells and other molluscan forms of African origin such as *Aspatharia cailliaudi, Unio abyssinicus,* etc. Included also are some late Acheulian implements in its upper levels.

Reference

Said *et al.* (1970).

ROSETTA EVAPORITES
Late Miocene (Tmu)

Type section

Rosetta offshore well no. 2 from depth 2678 to 2720 m. The coordinates of the well: lat. 31°37'22" N; long. 30°32'34" E.

Lithology

The formation is made up of layers of anhydrite interbedded with thin layers of clay.

Stratigraphic limits

Overlies the Qawasim and underlies the Abu Madi formation.

Extent

North Delta embayment in the subsurface.

Fossils and age

Nonfossiliferous, Messinian.

Reference

Rizzini *et al.* (1978).

SAHABA FORMATION
Late Pleistocene (Q₃)

Type section

Name proposed by de Heinzelin (1968) for the unit of massive silts and fine-grained micaceous sand and pebbles occurring as embankments 35 m higher than the flood plain level (prior to the erection of the Aswan High Dam). The type section is site 81 of the Combined Prehistoric Expedition, Sahaba district, Sudanese Nubia. It forms part of the Old Aggradational Silts of the Neonile. See Sahaba-Darau Formation.

SAHABA-DARAU FORMATION
Late Pleistocene (Q₃)

Type section
Pit 1, site E 61 M2, of the Combined Prehistoric Expedition, Dishna, east bank of Nile.

Lithology
Forms the γ silts of the older aggradational deposits of the Neonile. The unit is made up of a massive silt unit of light brownish gray color (2.5 Y 6/2) with slickensides and medium prismatic structure. In places it is made up of fine-grained micaceous sands and pebble beds of different composition.

Stratigraphic limits
In the type section the unit overlies a thin bed of dune sands (the Ikhtiariya) which overlies and interfingers a bed of gravel and coarse sand (the Korosko). On the west bank and near the trough of the Nile it overlies the Masmas-Ballana silts forming embankments about 12 m higher than the modern flood plain level. In the type section the unit is overlain by the Dishna playa deposits.

Extent
Along both banks of the Nile.

Fossils and age
Rich in molluscan remains, the most common of which are *Unio willcocksi* found in association with either the bivalve *Corbicula fluminalis* or the gastropod *Cleopatra bulimoides*. Mammalian faunas, generally found in human occupation sites, are rather abundant: Hartebeast (*Bubalis buselaphus*), gazelle (*Gazella dorcas isabella*), cow (*Bos primigenitus*), extinct buffalo (*Bubalis vignardi*), wild ass (*Equus asinus*), hippo (*Hippopotamus amphibius*), hyena (*Hyaena crocuta*), crocodile (*Crocodilus niloticus*), tortoise (*Testudo* sp.), ostrich, fishes, and others. Archeological materials and radiocarbon dates give the age of this unit between 15,000 and 12,000 B.P.

Remarks
The Sahaba-Darau was proposed by Wendorf and Schild, 1976 to include the Deir el-Fakhuri and the upper silts (Sahaba) of Said *et al.* (1970). In the present work the term is retained for the aggradational silts of the γ Neonile only. The name is composed of the Sahaba (Sudanese Nubia) and the Darau member of the Gebel Silsila formation of Butzer and Hansen (1968) in Kom Ombo.

Reference
Wendorf and Schild (1976).

SHATURMA FORMATION
Holocene
Name proposed by Butzer and Hansen (1968) for the wadi gravels and sands of modern aspect. Mapped in the accompanying geological maps as Fanglomerates and recent wadi alluvium (Qw).

SIDI SALEM FORMATION
Early and Middle Miocene (Tmm and Tml)

Type section
Sidi Salem well # 1 from depth 3685 to 4038 m (total depth). The coordinates of the well: lat. 31°19'10" N; long. 30°43'16" E.

Lithology
The formation is made up of clay with a few interbeds of sandstone, siltstone, and dolomite. The clays are gray-green and consist mainly of kaolinite with montmorillonite and small quantities of illite. The sandstones are quartzose and cemented by clay and calcite.

Stratigraphic limits
The lower limit is unknown. The upper limit is marked by the appearance of the sands of the overlying Qawasim formation. In the offshore area the Sidi Salem rests directly under the Rosetta evaporites.

Extent
North Delta embayment in the subsurface.

Fossils and age
The formation carries a rich open marine microfauna of the early and middle Miocene. The *Orbulina* datum which is taken by many authorities in Egypt to separate the early and middle Miocene cuts across the formation.

Reference
Rizzini *et al.* (1978).

SINQARI MEMBER
See Ineiba Formation

EL WASTANI FORMATION
Early and Middle Pleistocene?

Type section:
el-Wastani well # 1 from depth 1009 to 932 m. The coordinates of the well: lat. 31°24'08" N; long. 31°35'46" E.

Lithology

The formation is made up of thick sand beds alternating with thin clay beds. The sands are quartzose, coarse to medium-grained with little feldspars. The clays are sandy.

Stratigraphic limits

Overlies the Kafr el-Sheikh formation and underlies the coastal sands of the Mit Ghamr formation.

Fossils and age

A few brackish water forms were separated from the clay layers. The age is either middle or lower Pleistocene.

Extent

North Delta embayment in the subsurface.

Reference

Rizzini *et al.* (1978).

YOUNGER NEONILE DEPOSITS:
See Neonile Deposits

APPENDIX C

New Reflection Seismic Evidence of a Late Miocene Nile Canyon

by Felix P. Bentz[1] and Judson B. Hughes[2]

Reflection seismic data recorded in the Nile Delta area just north of Cairo by Santa Fe Minerals (Egypt), Inc., and its partners[3] provided new evidence of a late Miocene Nile canyon. This 2500-m deep gorge was cut by the river Nile as a result of the desiccation of the Mediterranean during Messinian time.

SEISMIC PROGRAM AND PROCESSING

In May 1975 Santa Fe and its partners began a reflection seismic program in the Nile delta north of Cairo as the initial effort in the exploration for oil and gas in the East Cairo Production Sharing Agreement Area extending from the Damietta branch of the Nile river to the Great Bitter Lake in the east. The area of interest covered by our discussions is shown on Figure 67. Western Geophysical Company was contracted to perform both seismic data gathering and data processing operations. A conventional CDP reflection seismic recording operation was conducted using 4 truck-mounted Vibroseis[4] sources with 48 channels of digitally recorded common-depth-point data.

[1] Vice President, Foreign and Offshore Exploration, Santa Fe Minerals Inc., Dallas.

[2] Chief Geophysicist of Santa Fe Minerals Inc., Dallas, Texas.

[3] Bow Valley Exploration (Egypt), Ltd.; Central Energy Development Co., Ltd.; and CITCO International Petroleum Company.

[4] Vibroseis is a registered trademark of the Continental Oil Co.

Due to the intensely farmed nature of delta lands, recording operations were necessarily restricted to existing roads and levees, which resulted in an irregular spacing and bearing of the lines (see Figure 68).

Since this was the first time that modern reflection seismic data had been recorded in this area, a thorough experimental program was undertaken to optimize signal quality and frequency range. The first lines were recorded with 40–10 Hz sweep frequency, 32 sweeps per VP with 150-m group spacing, 36 geophones/group in an in-line pattern (Lines A-40, A-45), and later tightened to 100-m spacing (Lines A-37, A-42, A-43, and A-44).

Initial results were very puzzling in the field and little sense could be made of the monitor records. At Western's London processing center an intensive effort was required to obtain coherent reflectors.

Lines A-43 and A-44 were two of the earliest lines processed and as can be seen by viewing Figure 70, we were startled to see an unexpectedly complex structural picture. It was soon evident, however, that a major unconformity underlain by a series of fault blocks could be interpreted. But how could this be deciphered geologically?

INTERPRETATION OF GEOPHYSICAL DATA

The first key to unraveling the general stratigraphy was a study of the seismic processing velocities obtained from the common-depth-point gathers. Interval velocity calculations

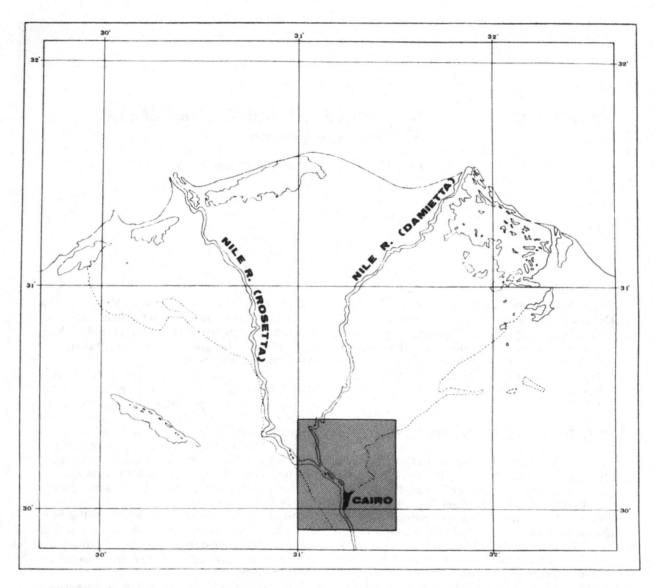

Figure 67. Area of interest shown in Figures 68 and 73.

from these RMS velocities revealed the following:

- The sub-unconformity blocks had interval velocities in the 3500–5000 m/sec range, suggesting carbonate rocks.

- The zone of coherent post-unconformity reflections lying between 1.2 and 1.8 sec had interval velocities in the 3000–3500 m/sec range, suggesting consolidated marine clastic sediments.

- The thickest overlying sequence of randomly reflecting material had interval velo-

cities in the 2000–2500 m/sec range, suggesting unconsolidated fluviatile deposits.

Following processing of all lines recorded in the delta, we prepared a contour map in reflection time of this unconformable surface (see Figure 68). In order to best illustrate the subsurface expression of the Nile canyon unconformity, we chose the lines shown in Figures 69–72. A time-to-depth conversion scale, derived from RMS velocity calculations is shown on the right-hand side of each section; however, it is applicable only for depth conversion to the unconformity surface. This time-to-depth conversion indicates that the Nile canyon was incised to a depth of about 2500 m

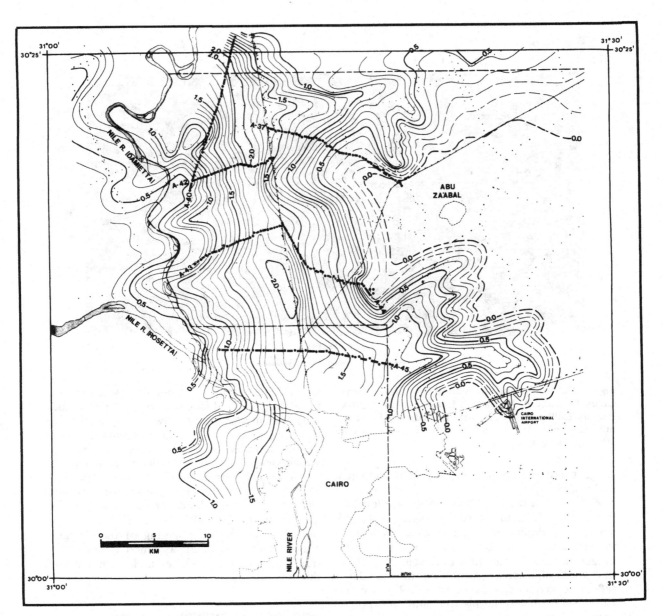

Figure 68. Late Miocene unconformity surface interpreted from reflection seismic sections. Contours in 2-way time, 10 msec interval.

below present sea level. One each figure we have marked the interpreted unconformity surface with a thin line; the inferred marine strata (later interpreted as early Pliocene marine sediments) lying in the bottom of the gorge have been shaded. The deeper pre-unconformity block faulting is clearly evident. The reader is invited to cross-reference continually with Figure 68.

Figure 69: Lines A-42 and A-37 located in the northern portion of our exploration area reveal the Nile canyon unconformity quite clearly. The strong reflector at 0.6 sec beneath SP 61 and continuing eastward is the upper Oligocene Abu Za'abal basalt, a thin sheet (30–50 m) which is widely spread throughout the Eastern Desert area as well as west of the Nile Valley. This reflector has been verified by a number of exploratory wells drilled in the Eastern Desert and can be tied directly to the surface type locality at Abu Za'abal, where the unweathered basalt is quarried. The age of the oldest rocks cut by the Nile River can only be inferred from velocity and seismic stratigraphic correlations. However, the results of our drilling campaign to the east of the delta strongly

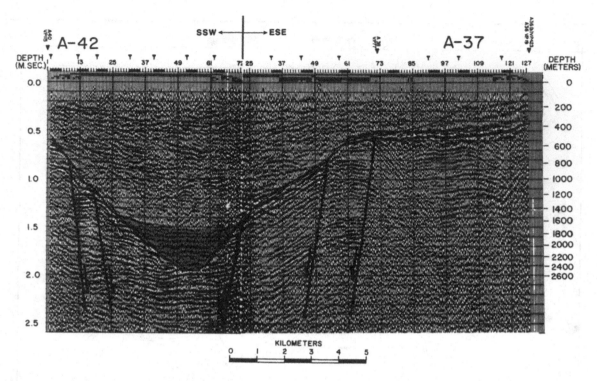

Figure 69. Composited reflection seismic sections, Lines A-42 and A-37. Base of Eonile canyon approximately 2500 meters deep. Interpreted early Pliocene infill has been shaded. Abu Za'abal basalt on right, 400-500 m deep.

suggest that the gorge bottomed in carbonates of Jurassic age.

Figure 70: Lines A-43 and A-44 are located further south than those in Figure 69 and suggest a widening of the gorge which is at least in part due to the orientation of Line A-44 tangentially to the east face of the canyon. Note the good character correlation in the deep reflective sequences between 1.9–2.3 sec on Line A-43 substantiating the fault interpretation.

Figure 71: Line A-45, although relatively poor in data quality (due to high noise levels through North Cairo traffic), illustrates the deep block faulting quite clearly. Here, the apparent widening of the gorge is actually caused by a subsidiary erosion channel entering the Nile canyon from the east.

Figure 72: Line A-40 illustrates the most northwesterly cut of our seismic data across the Nile canyon. The cuesta-like fault blocks on the left are thought to be capped by the highly resistant Mokattam limestones of middle Eocene age.

In addition to the few seismic lines repro-

duced here, numerous other lines were used in preparing the final isochron map of this unique erosional unconformity shown on Figure 68. The resulting picture of a major canyon north of Cairo reminded us of a Bouguer gravity map which we had received earlier from the Egyptian General Petroleum Corporation (EGPC). Figure 73 is a pertinent portion of this map showing the coincidence of the major gravity low north of Cairo with our seismic interpretation. Further depth modeling of the Bouguer gravity values with an assumed density contrast of 2.0 gm/cc produced a close match with our seismic unconformity.

GEOLOGICAL INTERPRETATION

The overwhelming geophysical evidence of a gigantic chasm underlying the fertile plain of the Nile delta demanded a compelling geological explanation. Almost immediately an article by Kenneth J. Hsu published in the December 1972 issue of *Scientific American* came to mind. Its title was "When the Mediterranean Dried Up" and it described the startling discoveries made by the scientists of the Glomar Challenger during Leg 13 of the Deep Sea Drilling Project.

At that time and for some years thereafter Hsu's advocation of a desiccated Mediterra-

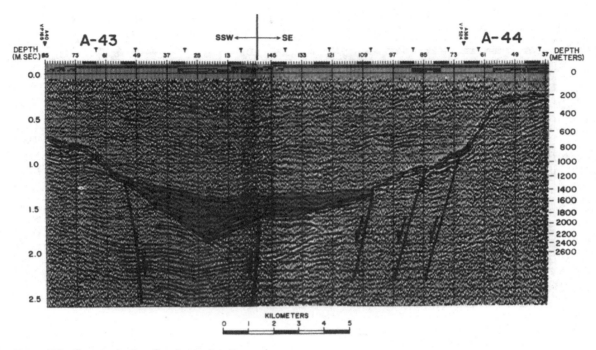

Figure 70. Composited reflection seismic sections, Line A-43 and A-44 across Eonile canyon. Influence of older block faulting is evident.

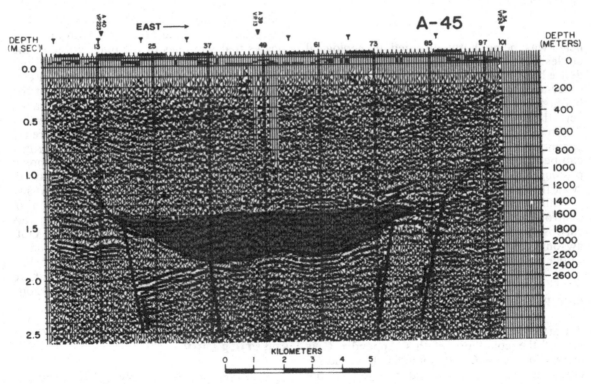

Figure 71. Reflection seismic section, Line A-45. Apparent broadening of canyon is due to tributary canyon entering from east.

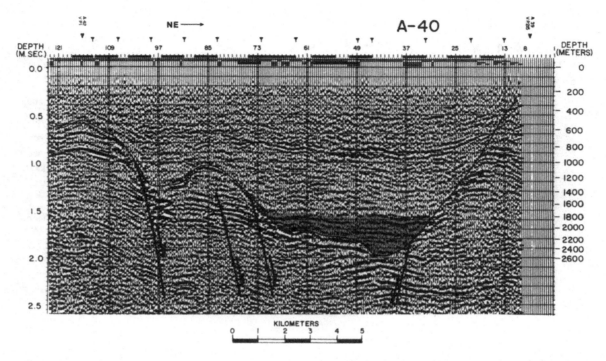

Figure 72. Reflection seismic section, Line A-40. Cuestas of middle Eocene Mokattam limestone probably cap fault blocks beneath unconformity on the left side of Eonile Canyon.

nean was being hotly disputed in geological circles. However, over the years sufficient evidence has been accumulated by Said (1975) and other authors quoted in the main body of this book, to make the "Messinian Event" an accepted reality.

Our geophysical evidence of an erosional channel cut 2500 m below the suburbs of Cairo can be regarded as the "Rosetta Stone" of the prolonged quest to decipher the Neogene history of the Mediterranean.

While relatively little can be added to the incontrovertible seismic evidence, a few geological observations may enhance the picture of the awsome gorge encised by the Eonile in the late Miocene landscape of Egypt.

MORPHOLOGY AND STRUCTURE

An artistic conception of the Nile canyon by one of the coauthors of this Appendix C is shown in Figure 64. It immediately evokes a comparison with the Grand Canyon of the Colorado River in Arizona. Said discusses this analogy at some length in his text (pp. 155-157). Another aspect, however, is worth mentioning,

namely, the apparent influence of fault patterns on the trend of these great canyons.

A geological map of the Bright Angel Quadrangle, published by John H. Maxon (1968 Edition) and supplemented by our own Landsat interpretations of the Grand Canyon area, indicates that stretches of the inner gorge of the Colorado River as well as a number of its tributaries are influenced by fault zones. Similar evidence is visible on the seismic sections of the Nile canyon, suggesting a pre-erosion graben complex controlled by northerly and northwesterly trending fault systems. These fault trends are most likely of Oligocene age and are probably related to the outpouring of the widespread Abu Za'abal basalt flows. The importance of this fault pattern as distinct linears had already been recognized during an interpretation of Landsat data conducted by Santa Fe Minerals prior to the seismic program (Bentz and Gutman, 1977). The continuation of this zone of crustal weakness in a northerly direction below the present Nile delta may also account, to some extent for the north–south trending zone of seismicity as shown on Figure 16.

STRATIGRAPHY

The stratigraphy of the Nile Valley and adjacent areas is treated in full detail by Said in the text of this volume. Additional geological information directly attributable to the explor-

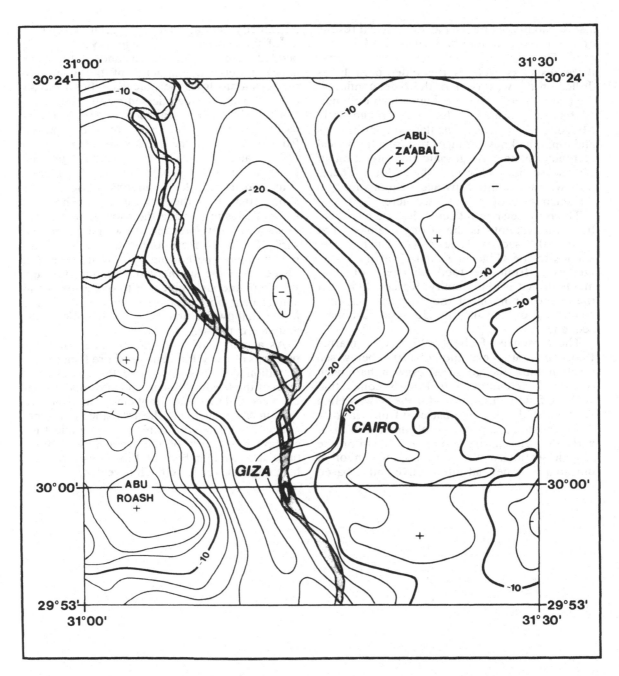

Figure 73. Bouguer gravity contour map showing major negative anomaly coincident with Eonile Canyon. Contour interval, 2 mgals.

atory efforts of the Santa Fe Group consists of a detailed photogeological map, supplemented by field observations covering the eastern rim of the Nile canyon area in the vicinity of Cairo. This mapping confirmed the previously established sequence of Neogene events:

(1) Early Miocene marine transgression over

the Oligocene Abu Za'abal basalts and older beds.

(2) The last major tectonic phase of folding, faulting and uplift.

(3) Late Miocene erosion resulting in widespread gravel deposits.

(4) Lower Pliocene marine evasion along the Nile Valley and its tributaries.

The absence of marine upper Miocene was

also substantiated by the paleontological results of the three exploratory wells drilled by the Santa Fe Group in the Eastern Desert. These three tests, as well as the older bore holes drilled on the west bank of the Nile, confirmed the highly mobile shelf geology which dominated this area from late Jurassic through early Miocene. Numerous angular unconformities and rapid thickness changes resulting from local tectonics make stratigraphic interpretation of seismic sections extremely hazardous. Therefore, we have omitted any age identifications on the seismic sections presented here.

The only seismic reflector that we can identify with certainty is the strong event seen at around 0.6 sec on Line A-37, which corresponds to the Oligocene Abu Za'abal basalts. However, from the total thickness of sediments drilled in the vicinity, we conclude that the deepest portion of the Nile canyon must have been eroded into the Jurassic carbonate sequence.

The presence of about 120 m of a marine Pliocene fill at an elevation of 170 m below sea level in the Nile gorge at Aswan had been reported by Chumakow in 1967 in connection with the foundation drilling for the Aswan High Dam located 1000 km upstream from Cairo.

The postulated presence of about 800 m of early Pliocene marine sediments shown on the seismic sections at the bottom of the canyon in our area is somewhat interpretive and is based on velocity data as well as the relative regularity of the seismic events suggestive of evenly bedded marine sediments. In contrast, the more irregular seismic character of the overlying events and the lower seismic velocities are typical of fluviatile to deltaic sediments.

There is no inconsistency in the fact that early Pliocene oyster beds are also reported from outcrops in the Heliopolis area east of Cairo (Said, 1962a). One can easily imagine the Pliocene sea rushing back into the Mediterranean and up the deep canyons incised on its shores and filling them to the brim. The sediments deposited during the short time interval of such an early Pliocene sea would not necessarily fill these deep gorges to the brim, but mainly cover the bottom and lay down a thin veneer over the submerged promontories, and yet simultaneously give rise to oyster beds and similar nearshore deposits along the coasts and in the shallower embayments of the tributaries and side canyons.

As a matter of fact, the presence of these outcrops of early Pliocene seashore facies at an elevation only slightly above the current sea level of the Mediterranean indicates that probably very little isostatic movement in either direction has occurred in the area since that time.

Had it not been for man's unending quest for energy, the secret of the tremendous forces once at work deep beneath this now placid surface might never have been revealed.

References

Abdallah, A. H., and el-Kadi, M., 1974. Gypsum deposits in the Faiyum province. *Fac. Sci., Cairo Univ. Bull.* **47**, 335–345.

Albritton, C. C., 1968. Geology of the Tushka Site. In: *The Prehistory of Nubia*, 1, Wendorf, F. (ed.). Southern Methodist University Press, Dallas, Texas, 856–864

Aleem, A. A., 1972. Effect of river outflow on marine life. *Marine Biol.* **15**, 200–208.

Andrew, G., 1948. Geology of the Sudan. In: *Agriculture in the Sudan*, Tothill, J. D. (ed.). Government Press, Khartoum, 84–128.

Andrews, C. W., 1902. Note on a Pliocene Vertebrate Fauna from the Wadi Natrun, Egypt. *Geol. Mag. (London)* **9**, 433–439.

Andrews, C. W., 1906. *The Extinct Animals of Egypt.* A descriptive Catalogue of the Tertiary Vertebrates of the Faiyum, Egypt, based on the collection of the Egyptian Government in the Geological Museum (Natural History), London. London, 324 pp.

Anwar, Y. M., and Khadr, M., 1958. Mineralogical study of the soils of Kharga Oasis. *Fac. Sci. Alexandria Univ., Bull.* **2**, 138–159

Arkell, A. J., 1949. *The Old Stone Age in the Anglo-Egyptian Sudan.* Sudan Antiquities Service, Occasional Papers 1, Khartoum, 52 pp.

Arldt, T., 1911. Entstehung des Niltals. *Naturwiss. Rundschau Braunsweig* **26**, 27–41.

Arrhenius, G., 1952. *Sediment cores from the East Pacific.* Swedish Deep Sea Exped. (1947–1948) Reports, 5(1), 89 pp.

Attia, M. I., 1954. *Deposits in the Nile Valley and the delta.* Mines and Quarries Dept., Geol. Surv. Egypt, Cairo, 356 pp.

Awad, M. 1928. Some stages in the evolution of the River Nile. *Proc. Internat. Geogr. Congr.,* Cambridge, 267–288.

Ball, J., 1939. *Contributions to the geography of Egypt.* Survey and Mines Dept., Cairo, 308 pp.

Barr, F. T., and Walker, B. R., 1973. Late Tertiary channel system in northern Libya and its implications on Mediterranean sea level changes. In: *Initial Reports of the Deep Sea Drilling Project,* Leg XIII, 2, Ryan, W. B. F., and Hsu, K. J. (eds.). U.S. Government Printing Office, Washington, D.C., 1244–1251.

Beadnell, H. J. L., 1901. *Some recent geological discoveries in the Nile Valley and Libyan Desert.* (An English translation of a paper communicated to the Inter. Geol. Congr., Paris, 1900). Stephen Austin and Son, London, 24 pp.

Beadnell, H. J. L., 1905. *The topography and geology of the Faiyum Province of Egypt.* Survey Dept. Egypt, 101 pp.

Bell, B., 1970. The oldest records of the Nile floods. *Geogr. J.* **136**, 569–573.

Bell, B., 1971. The dark ages in ancient history, *Amer. J. Archaeol.* **75**, 1–26.

Bellini, E., 1969. Biostratigraphy of the Al Jaghbub (Giarabub) Formation in Eastern Cyrenaica (Libya). *Proc. 3rd African Micropal. Colloquim, Cairo,* 165–184

Bentz, F. P., and Gutman, S. J., 1977. Landsat data contributions to hydrocarbon exploration in foreign regions. *U.S. Geol. Surv. Prof. Paper* **1015**, 83–92.

Berggren, W. A., 1969. Rates of evolution in some Cenozoic planktonic foraminifera. *Micropaleontology* **15**, 351–369.

Berggren, W. A., 1971. Neogene chronostratigraphy, planktonic foraminiferal zonation and the radiometric time-scale. *Hung. Geol. Soc. Bull.* **101**, 162–169.

Berggren, W. A., 1972. A Cenozoic time-scale, some implications for regional geology and paleobiogeography. *Lethaia* **5**, 195–215.

Berggren, W. A., and Van Couvering, J. A., 1974. *The Late Neogene.* Elsevier, Amsterdam and New York, 216 pp.

Berggren, W. A., Phillips, J. D., Bertels, A., and Wall, D., 1967. Late Pliocene-Pleistocene stratigraphy in deep-sea cores from the south-central North Atlantic. *Nature* **216**, 253–254.

Biberson, P., Coque, R., and Debono, F., 1977. Decouverte d'industries preacheuleennes *in situ* dans les formations du piemont de la montagne de Thebes. *C. R. Acad. Sci. (Paris)* **285**(D), 303–305.

Bietak, M., 1974. Tell el-Dab'a II, Der Fundort im Rahmen einer archaeologisch-geographischen Untersuchung uber das ägyptische Ostdelta. *Osterreich. Akad. Wiss., Denkschr. Gesamakad.* **4**, 1–236.

Blanckenhorn, M. L. P., 1901. Neues zur Geologie und Palaeontologie Aegyptens IV, Das Pliocaen- und Quartaerzeitalter in Aegypten ausschliesslich des Rothen-Meergebietes, *Zeitschr. D. Geol. Ges. (Berlin)* **53**, 307–502.

Blanckenhorn, M. L. P., 1910. Neues zur Geologie Palastinas und des aegyptischen Niltals. *Zeitschr. D. Geol. Ges. (Berlin)* **62**, 405–461.

Blanckenhorn, M. L. P., 1921. *Handbuch der Regionalen Geologie: Aegyptens.* Heidelberg, 244 pp.

Bovier-Lapierre, P., 1926. Les gisements paleolithiques de la plaine de l'Abbassieh. *Inst. Egypte. Bull.* **8**, 257–275.

Brooks, C. E. P., 1949. *Climate through the ages.* McGraw Hill, New York, 395 pp.

Butzer, K. W., 1959a. Contributions to the Pleistocene geology of the Nile Valley. *Erdkunde* **13**, 46–47

Butzer, K. W., 1959b. Die Naturlandschaft Aegyptens waehrend der Vorgeschichte und dem dynastichen Zeitalter. *Akad. Wiss. Lit. Math.-Nat. Kl. (Mainz)* **12**, 1–81.

Butzer, K. W., 1960a. On the Pleistocene shore lines of Arabs' Gulf, Egypt. *J. Geol.,* **68**, 626–637.

Butzer, K. W., 1960b. Archeology and geology in ancient Egypt. *Science* **132**, 1617–1624.

Butzer, K. W., 1964. *Environment and Archeology.* Aldine, Chicago, 524 pp.

Butzer, K. W., and Hansen, C. L., 1965. Upper Pleistocene stratigraphy in southern Egypt. In: *Symposium on Background to Evolution in Africa,* Bishop, W. W. and Clark, J. D., (eds.). Austria, 329–356.

Butzer, K. W., and Hansen, C. L., 1968. *Desert and River in Nubia.* University of Wisconsin Press, Madison, 562 pp.

Butzer, K. W., Isaac, G. L., Richardson, J. L., and Washbourn-Kamau, C. K., 1972. Radiocarbon dating of East African lake levels. *Science* **175**, 1069–1076.

Caton-Thompson, G., and Gardner, E. W., 1934. *The Desert Faiyum.* Royal Anthrol. Inst., London, 2 vols.

Chatters, R. M., 1968. Washington State University natural radiocarbon measurements I. *Radiocarbon* **10**, 479–498.

Chavaillon, J., 1964. *Etude Stratigraphique des Formations Quaternaires du Sahara Nordoccidental.* Cent. Nat. Rech. Sci., Paris, 393 pp.

Chumakov, I. S., 1967. Pliocene and Pleistocene deposits of the Nile Valley in Nubia and Upper Egypt. (In Russian). *Trans. Geol. Inst. Acad. Sci. USSR* **170**, 1–110.

Churcher, C. S., 1972. *Late Pleistocene vertebrates from archaeological sites in the plain of Kom Ombo, Upper Egypt.* Life Sci. Contr., Roy. Ontario Mus. **82**, 172 pp.

Churcher, C. S., 1974. Relationships of the Late Pleistocene Vertebrate fauna from Kom Ombo, Upper Egypt. *Geol. Surv. Egypt, Annals* **4**, 363–384.

Churcher, C. S., and Smith, P. E. L., 1972. Preliminary report on the fauna of Late Paleolithic sites in Upper Egypt. *Science* **177**, 259–261.

Ciaranfi, N., and Cita, M. B., 1973. Paleontological evidence of changes in the Pliocene climates. In: *Initial Reports of the Deep Sea Drilling Project* Leg XIII, 2, Ryan, W. B. F., and Hsu, K. J. (eds.). U. S. Government Printing Office, Washington, D. C., 1387–1399.

Cita, M. B., 1973a. Inventory of biostratigraphical findings and problems. In: *Initial Reports of the Deep Sea Drilling Project,* Leg XIII, 2, Ryan, W. B. F., and Hsu, K. J. (eds.). U. S. Government Printing Office, Washington, D. C., 1045–1073.

Cita, M. B., 1973b. Pliocene biostratigraphy and chronostratigraphy. In: *Initial Reports of the Deep Sea Drilling Project,* Leg XIII, 2, Ryan, W. B. F., and Hsu, K. J. (eds.). U. S. Government Printing Office, Washington, D. C., 1343–1397.

Cita, M. B., and Ryan, W. B. F., 1973. Time-scale and general synthesis. In: *Initial Reports of the*

Deep Sea Drilling Project, Leg XIII, 2, Ryan, W. B. F., and Hsu, K. J. (eds.). U. S. Government Printing Office, Washington, D. C., 1405–1416.

Coque, R., 1962. *La Tunisie Presaharienne: Etude Géomorphologique.* A. Colin, Paris, 476 pp.

Coque, R., and Said, R., 1972. Observations preliminaire sur la géomorphologie et le Quaternaire de Piemont Thebaine: *Grafitti de la Montagne Thebaine,* 1. Centre de Documentation et Etudes sur l'ancienne Egypte, Le Caire, 12–22.

de Heinzelin, J., 1968. Geological history of the Nile Valley in Nubia. In: *The Prehistory of Nubia,* 1, Wendorf, F. (ed.). Southern Methodist University Press, Dallas, Texas, 19–55.

de Heinzelin, J., and Paepe, R., 1965. The geological history of the Nile Valley in Sudanese Nubia: Preliminary Results. In: *Contr. to the Prehistory Nubia,* Wendorf, F. (ed.). Southern Methodist University Press, Dallas, Texas, 29–56.

el-Gabaly, M. M., and Khadr, M., 1962. Clay mineral studies of some Egyptian desert and Nile alluvial soils. *J. Soil Sci.* **13,** 33–342.

el-Wakeel, S., and el-Sayed, M. K., 1978. The texture, mineralogy and chemistry of bottom sediments and beach sands from the Alexandria region, Egypt. *Marine Geol.* **27,** 137–160.

Emery, K. O., and Bentor, Y. K., 1960. The continental shelf of Israel. *Geol. Surv. Isreal, Bull.* **26,** 25–41.

Emery, K. O., Heezen, B. C., and Allan, T. D., 1966. Bathymetry of the eastern Mediterranean Sea. *Deep Sea Res.* **13,** 173–192.

Emiliani, C., 1966. Paleotemperature analysis of the Caribbean cores P 6304-8 and P 6304-9 and a generalized temperature curve for the last 425,000 years. *J. Geol.* **74**(6), 109–126.

Ericson, D. B., and Wolling, G., 1968. Pleistocene climates and chronology in deep-sea sediments. *Science* **162,** 1227–1234.

Evans, P., 1972. The present status of age determination in the Quaternary (with special reference to the period between 70,000 and 1,000,000 years ago). *24th Intern. Geol. Congr.,* Canada, Section 12, 16–21.

Fairbridge, R. W., 1962. New radio-carbon dates of Nile sediments. *Nature* **196,** 108–110.

Fairbridge, R. W., 1963. Mean sea level related to solar radiation during the last 20,000 years. In: *Symposium on changes of climate. Proc. Unesco and the World Meteorological Organization,* 229–240.

Fairbridge, R. W., 1967. Global climate change during the 13,500 B. P. Gothenburg geomagnetic excursion. *Nature* **265,** 430–431.

Farag, I. A. M., and Ismail, M. M. 1959. Contributions to the stratigraphy of the Wadi Hof area (north-east of Helwan). *Fac. Sci., Cairo Univ. Bull.* **34,** 147–168.

Flint, R. F., 1959. Pleistocene climates in Eastern and Southern Africa. *Geol. Soc. Amer. Bull.* **70,** 343–374.

Fourtau, R., 1915. Contributions a l'etude des Dépots nilotiques. *Inst. Egypte Mem.* **8,** 57–94.

Galanopoulos, A. G., 1970. Plate tectonics in the area of Greece as reflected in the deep focus seismicity. *Geol. Soc. Greece Bull.* **10,** 67–71.

Geological Map of Egypt, 1928. In: *Atlas of Egypt,* Egypt Survey Dept.

Gergawi, A., and Khashab, H. M. A., 1968a. Seismicity of U. A. R. *Helwan Obs. Bull.* **76,** 1–27.

Gergawi, A., and Khashab, H. M. A., 1968b. Seismic wave velocities in U. A. R. *Helwan Obs. Bull.* **77,** 1–25

Giegenagack, R. F., 1968. Late Pleistocene history of the Nile Valley in Egyptian Nubia. Ph.D. Dissertation, Yale University (unpublished).

Gorshkov, G. P., 1963. The seismicity of Africa. In: *A review of the natural resources of the African continent.* UNESCO, Paris, 101–105.

Gregory, J. W., 1920. The African rift valleys. *Geogr. J.* **56,** 13–47; 327–328.

Gribbin, J., and Lamb, H. H., 1978. Climatic change in historical times. In: *Climatic Change,* Gribbin, J. (ed.). Cambridge University Press, 68–82.

Gvirtzman, G., 1969. The Saqiye Group (Late Eocene to Early Pleistocene) in the coastal plain and Hashephela region, Israel. *Geol. Survey Israel* Rep. OD/5/67 (Quoted in Neev, 1969).

Gvirtzman, G., and Buchbinder, B., 1977. The desiccation events in the Eastern Mediterranean during Messinian times and compared with other Miocene desiccation events in basins around the Mediterranean. In: *Internat. Symp. Structural History Mediterranean Basins,* Split, Yugoslavia, Biju-Duval, B., and Montadert, L. (eds.). Edition Technip, Paris, 411–420.

Hachett, J. P., and Bischoff, J. L., 1973. New data on the stratigraphy, extent and geologic history of the Red Sea geothermal deposit. *Econ. Geol.* **68,** 553–564.

Hamdi, H., 1967. The mineralogy of the fine fraction of the alluvial soils of Egypt. *U. A. R. J. Soil Sci.* **7,** 15–21.

Harrison, J. C., 1955. An interpretation of gravity anomalies in eastern Mediterranean. *Phil. Trans. Roy. Soc.* A **248,** 283–325.

Hassan, F., 1974. *The archeology of the Dishna Plain, Egypt: A Study of a Late Paleolithic Settlement.* Geol. Surv. Egypt, Paper 59, 174 pp.

Hassan, F., 1976. Heavy minerals and the evolution of the modern Nile. *Quat. Res.* **6**, 425–555.

Hayes, W. C., 1964. *Most Ancient Egypt.* Chicago University Press, 160 pp.

Hays, J. D., 1965. Radiolara and Late Tertiary and Quaternary History of Antarctic Seas. In: *Biology of the Antarctic Seas* II. Antarctic Research Amer. Geophys. Union **5**, 125–184.

Hays, J. D., Saito, T., Obdyke, N. D., and Burckle, L. H., 1969. Pliocene-Pleistocene sediments of the Equatorial Pacific: Their paleomagnetic biostratigraphic and climatic record. *Geol. Soc. Amer. Bull.* **80**, 1481–1513.

Holmes, A., 1965. *Principles of Physical Geology.* Ronald Press, New York, 1288 pp.

Hsu, K. J., 1972. When the Mediterranean dried up. *Scientific Amer.* December, 27–36.

Hsu, K. J., and Cita, M. B., 1973. The origin of the Mediterranean Evaporite. In: *Initial reports of the Deep Sea Drilling Project,* Leg XIII, 2, Ryan, W. B. F., and Hsu, K. J. (eds.). U. S. Government Printing Office, Washington, D. C., 1203–1232.

Hsu, K. J., and Montadert, L., (eds.), 1977. *Initial Reports of the Deep Sea Drilling Project* **42**. U. S. Government Printing Office, Washington, D. C.

Hsu, K. J., and Ryan, W. B. F., 1973. Summary of the evidence for extensional and compressional tectonics in the Mediterranean. In: *Initial reports of the Deep Sea Drilling Project,* Leg XIII, 2, Ryan, W. B. F., and Hsu, K. J. (eds.). U. S. Government Printing Office, Washington, D. C., 1011–1019.

Hsu, K. J., Ryan, W. B. F., and Cita, M. B., 1973. Late Miocene desiccation of the Mediterranean. *Nature* **242**, 239–243.

Hull, E. G., 1896. Observations on the geology of the Nile Valley and the evidence of the greater volume of the river at a former period. *Quart. J. Geol. Soc. (London)* **52**, 308–319.

Hunting Geology and Geophysics, 1967. *Assessment of the mineral potential of the Aswan region,* U. A. R.: photogeological survery. Unpublished report. Hunting Geology and Geophysics, Ltd., England, U. N. Development Program and U. A. R. Regional Planning of Aswan. 138 pp.

Hurst, H. E., 1944. *A short account of the Nile basin.* Physical Dept., Ministry of Public Works, Egypt, Paper 45, 77 pp.

Hurst, H. E., *et al.,* 1931-1966. *The Nile Basin.* Physical and Nile Control Dept., Ministry of Irrigation, Cairo, Vols. I–X.

Huzzayin, S. A., 1941. *The Place of Egypt in Prehistory.* Inst. Egypte Mem. **43**, 440 pp.

Irwin, H. T., Wheat, J. B., and Irwin, L. F., 1968. *University of Colorado Investigations of Paleolithic and Epipaleolithic sites in the Sudan, Africa,* Univ. Utah Papers in Anthropology, 90. University of Utah Press, Salt Lake City.

Ismail, A., 1960. Near and local earthquakes at Helwan from 1903–1950, *Welwan Obs. Bull.* **49**.

James, C. T., and Slaughter, B. H., 1974. A primitive new Middle Pliocene Murid from Wadi el Natrun, Egypt. *Geol. Surv. Egypt, Annals* **4**, 33–362.

Jarvis, C. S., 1936. Flood-stage records of the River Nile. *Amer. Soc. Civil Engineers, Trans.,* 101, 1012–1071.

Judd, J. W., 1897. Report on a series of specimens of the deposits of the Nile Delta, obtained by boring operations undertaken by the Royal Society, 2nd report. *Proc. Roy. Soc. (London)* **61**, 32–40.

Kendall, R. L., 1969. An ecological history of the Lake Victoria Basin. *Ecol. Monogr.* **39**, 121–167.

Kenyon, N. H., Stride, A. H., and Belderson, R. H., 1975. Plan views of active faults and other features on the lower Nile cone. *Geol. Soc. Amer. Bull.* **86**, 1733–1739.

Kholeif, M. M., 1973. Microscopic representation of the diatomite samples, Faiyum, Egypt. *7th Arab Sci. Congr. (Cairo)* **4**, 17–26.

Kholeif, M. M., Hilmy, M. E., and Shahat, A., 1969. Geological and mineralogical studies of some sand deposits in the Nile delta. *J. Sed. Petrology* **39**, 1521–1529.

Kukla, G. J., 1977. Pleistocene land-sea correlations, I. Europe. *Earth Sci. Rev.* **13**, 307–374.

Lamb, J. L., 1969. Planktonic foraminiferal datums and Late Neogene Epoch boundaries in the Mediterranean, Caribbean, and Gulf of Mexico. *Trans. Gulf Coast Assoc. Geol. Soc.,* **19**, 559.

Lawson, A. C., 1927. The Valley of the Nile. *Univ. Calif. Chronicles (Berkeley).* **29**, 235–259.

Le Pichon, S., Francheteau, J., and Bonnin, J., 1973. *Plate Tectonics, Developments in Geotectonics* **6**, Elsevier, Amsterdam and New York.

Little, O. H., 1935. Recent geological work in the Faiyum and in the adjoining portion of the Nile Valley. *Inst. Egypte Bull.* **18**, 201–240.

Lubell, D., 1974. *The Fakhurian, a Late Paleolithic industry from Upper Egypt.* Geol. Surv. Egypt, Paper 58, 193 pp.

Lyons, H. G., 1906. *The physiography of the River Nile and its basin.* Egypt Survey Dept., 411 pp.

Maldonado, A., and Stanley, D. J., 1976. The Nile

cone: submarine fan development by cyclic sedimentation. *Marine Geol.* **20**, 27–40.

Maley, J., 1970. Introduction a la géologie des environs de la deuxieme cataracte du Nil au Soudan. In: *Mirgissa I, Mission Archeologique Francaise au Soudan*, Vercoutter, J. (ed.), Paul Geuthner, Paris, 122–157.

Mansour, A. T., Barakat, M. G., and Abdel Hady, Y. S., 1969. Marine Pliocene planktonic foraminiferal zonation southeast of Salum, Egypt. *Riv. Ital. Paleont.* **75**(4), 833–842.

Marks, A. E., 1968. The Khormusan: An Upper Pleistocene industry in Sudanese Nubia. In: *The Prehistory of Nubia*, 1, Wendorf, F. (ed.). Southern Methodist University Press, Dallas, Texas, 315–391.

Maxon, J. H., 1968. Geodetic map of the Bright Angel Quadrangle, Grand Canyon National Park, Arizona. Grand Canyon Natural History Association, Flagstaff, Ariz.

Mayer-Eymar, C. D. W., 1898. Systematisches Verzeichniss der Fauna des unteren Saharianum (marines Quaterner) der Umgegend von Kairo, nebst Beschereibung der neuen Arten. *Palaeontographica* **30**, 60–90.

Meneisy, M. Y., and Kreuzer, H., 1974. Potassium-Argon ages of Egyptian basaltic rocks. *Geol. Jb.* **9**, 21–31.

Milliman, J. D., and Emery, K. E., 1968. Sea levels during the past 35,000 years. *Science* **162**, 1121–1123.

Misdorp, R., and Sestini, G., 1976a. The Nile Delta: Main features of the continental shelf topography. *Proc. Seminar Sedimentology Nile Delta, UNESCO-UNDP, Alexandria*, 145–161.

Misdorp, R., and Sestini, G., 1976b. Notes on a sediment map of the Nile Delta continental shelf, based on the Endeavour survey of 1919–1922. *Proc. Seminar Sedimentology Nile Delta, UNESCO-UNDP, Alexandria*, 191–204.

Neev, D., 1968. Submarine geological studies on the continental shelf and slope off the Mediterranean coast of Israel. *Israel Jour. Earth Sci.* **15**, 170–175.

Neev, D., 1975. Tectonic evolution of the Middle East and the Levantine basin. (Easternmost Mediterranean). *Geology* **3**, 683–686.

Neev, D., 1976. The geology of the southeastern Mediterranean Sea. *Geol. Surv. Israel Bull.* **68**, 1–51.

Neev, D., and Friedman, G., 1978. Late Holocene tectonic activity along the margins of the Sinai subplate. *Science* **203**, 131–142.

Nesteroff, W. D., 1973. The sedimentary history of the Mediterranean area during the Neogene. In: *Initial reports of the Deep Sea Drilling Project, Leg XIII*, 2, Ryan, W. B. F., and Hsu, K. J., (eds.). U.S. Government Printing Office, Washington, D.C., 1257–1262.

Omara, S., and Ouda, K., 1969. Pliocene foraminifera from the sub-surface rocks of Burg el Arab well no. 1, Western Desert, Egypt. *Proc. 3rd African Micropal. Colloq., Cairo*, 581–665.

Pfannenstiel, M., 1953. Das Quaternaer der Levante, II, Die Entstehung der aegyptischen Oasendepressionen. *Akad. Wiss. Lit. Math. Nat. Kl., Mainz*,

Phillip, G., and Yousri, F., 1964. Mineral composition of some Nile delta sediments near Cairo. *Fac. Sci. Cairo Univ. Bull.* **39**, 231–252.

Phillips, J. L., 1973. *Two final Paleolithic sites in the Nile Valley and their external relations*. Geol. Surv. Egypt, Paper 57, 110 pp.

Popper, W., 1951. *The Cairo Nilometer*. Univ. California Press.

Quenell, A. M., 1958. The structural and geomorphic evolution of the Dead Sea rift. *Quart. J. Geol. Soc. (London)* **114**, 1–24.

Richmond, G. M., 1970. Comparison of the Quaternary stratigraphy of the Alps and rocky mountains. *Quat. Res.* **1**, 3–28.

Rizzini, A., Vezzani, F., Cococcetta, V., and Milad, G., 1978. Stratigraphy and sedimentation of a Neogene-Quaternary section in the Nile Delta area. *Marine Geol.* **27**, 327–348.

Ross, D. A., and Schlee, 1973. Shallow structure and geologic development of the southern Red Sea. *Geol. Soc. Amer. Bull.* **84**, 3827–3848.

Ross, D. A., and Uchupi, E., 1977. Structure and sedimentary history of southeastern Mediterranean Sea—Nile cone area. *Amer. Assoc. Petrol. Geol. Bull.* **61**, 872–902.

Ryan, W. B. F., 1973. Paleomagnetic stratigraphy. In: *Initial Reports of the Deep Sea Drilling Project*, Leg XIII, 2, Ryan, W. B. F., and Hsu, K. H. (eds.). U.S. Government Printing Office, Washington, D.C., 1380–1386.

Ryan, W. B. F., 1976. Quantitative evaluation of the depth of the Western Mediterranean before, during and after the Late Miocene salinity crisis. *Sedimentology* **23**, 791–813.

Ryan, W. B. F., and Cita, M. B., 1978. The nature and distribution of Messinian erosional surfaces—Indicators of a several kilometer deep Mediterranean in the Miocene. *Marine Geol.* **27**, 193–230.

Ryan, W. B. F., *et al.* (eds.), 1973a. *Initial Reports of the Deep Sea Drilling Project*, Leg XIII, 2, U.S.

Government Printing Office, Washington, D.C., 515–1441.

Ryan, W. B. F., Venkatarathnam, K., and Wezel, F. C., 1973b. Mineralogical composition of the Nile cone, Mediterranean ridge and Strabo trench sandstones and clays. In: *Initial Reports of the Deep Sea Drilling Project,* Leg XIII, 2, Ryan, W. B. F., and Hsu, K. L. (eds.). U.S Government Printing Office, Washington, D.C., 731–746.

Rzoska, J. (ed.), 1976. *The Nile: Biology of an ancient river.* Monographiae Biologicae, **29.** Junk, The Hague, 417 pp.

Saad, S. I., and Sami, S., 1976. Studies of pollens and spores content of Nile Delta deposits (Berenbal region). *Pollen and Spores* **9,** 467–503.

Sabatini, R. R., Rabchevsky, G. A., and Sissala, J. E., 1971. *Nimbus Earth resources observations:* Tech. Rep. 2, NASA, Goddard Space Flight Center, Greenbelt, Maryland, 256 pp.

Said, R., 1955. Foraminifera from some "Pliocene" rocks of Egypt. *J. Washington Acad. Sci.* **45,** 8–13.

Said, R., 1958. Remarks on the geomorphology of the deltaic coast between Rosetta and Port Said. *Soc. Geogr. Egypte Bull.* **31,** 115–125.

Said, R., 1962a. *The Geology of Egypt.* Elsevier, Amsterdam and New York, 377 pp.

Said, R., 1962b. Das Miozaen in der westlichen Wueste Aegyptens. *Geol. Jb.* **80,** 349–366.

Said, R., 1971. *Explanatory Notes to accompany the geological map of Egypt.* Geol. Surv. Egypt, Paper 56, 123 pp.

Said, R., 1973. *Subsurface geology of Cairo Area.* Inst. Egypte Mem. **60.**

Said, R., 1974. Some observations on the geomorphology of the south Western Desert of Egypt. *Geol. Surv. Egypt, Annals* **5,** 61–70.

Said, R., 1975. The geological evolution of the River Nile. In: *Problems in Prehistory Northern Africa and the Levant,* Wendorf, F., and Marks, A. E. (eds.). Southern Methodist University Press, Dallas, Texas, 1–44.

Said, R., 1979. The Messinian in Egypt, *Proc. Intern. Congr. Mediterranean Neogene, Athens. Ann. Geol. Pays Helleniques, Hors ser., Fasc. III* 1083–1090.

Said, R., and Issawi, B., 1964. Preliminary results of a geological expedition to lower Nubia and to Kurkur and Dungal oasis. In: *Contribution to Prehistory of Nubia,* Wendorf, F. (ed.). Southern University Press, Dallas, 1–20.

Said, R., and Issawi, B., 1965. Geology of northern plateau, Bahariya Oasis, Egypt. *Geol. Surv. Egypt,* Paper 29, 1–41.

Said, R., and Yousri, F., 1964. Origin and Pleistocene history of River Nile near Cairo, Egypt. *Inst. Egypte Bull.* **45,** 1–30.

Said, R., Wendorf, F., and Schild, R., 1970. The geology and prehistory of the Nile Valley in Upper Egypt. *Archeologia Polana* **12,** 43–60.

Said, R., Wendorf, F., Albritton, G., Schild, R., and Kobusiewicz, M., 1972a. Remarks on the Holocene geology and archaeology of Northern Faiyum Desert. *Archaeologia Polana* **13,** 7–22.

Said, R., Wendorf, F., Albritton, G., Schild, R., and Kobusiewicz, M., 1972b. A preliminary report on the Holocene geology and archaeology of the Northern Faiyum Desert. In: *Playa Lake Symposium,* C. C. Reeves (ed.). Intern. Center Arid and Semi-arid Land Studies, Lubbock, Texas. Publ. 4, 41–61.

Salem, R., 1976. Evolution of Eocene-Miocene sedimentation patterns in parts of northern Egypt. *Amer. Assoc. Petrol. Geol. Bull.* **60,** 34–64.

Sandford, K. S., 1934. Paleolithic man and the Nile Valley in Upper and Middle Egypt. *Chicago University Oriental Inst. Publ.* **3,** 1–131.

Sandford, K. S., and Arkell, W. J., 1929. Paleolithic man and the Nile-Faiyum divide. *Chicago University Oriental Inst. Publ.* **1,** 1–77.

Sandford, K. S., and Arkell, W. J., 1933. Paleolithic man and the Nile Valley in Nubia, and Upper Egypt. *Chicago University Oriental Inst. Publ.* **17,** 1–92.

Sandford, K. S., and Arkell, W. J., 1939. Paleolithic man and the Nile Valley in Nubia and Upper Egypt. *Chicago University Oriental Inst. Publ.* **36,** 1–105.

Schwartz, H., Goldberg, G. P., and Marks, A. E., 1979. Uranium-series dating of travertine from archeological sites, Nahal Zin, Israel. *Nature* **277,** 558–560.

Servant, M., 1973. *Sequences Continentales et Variations Climatiques: Evolution du Bassin du Chad au Cenozoique Superieur.* Orstom, Paris.

Sestini, G., 1976. Geomorphology of the Nile Delta. *Proc. Seminar Sedimetology Nile Delta,* Alexandria, 12–24.

Shafei, A., 1940. Faiyum irrigation as described by Nabulsi in A.D. 1245 with a description of the present system of irrigation and a note on lake Moeris. *Inst. d'Egypte, Bull.* **20,** 283–327.

Shepard, F. P., 1964. *Submarine geology,* 2nd ed. Harper, New York.

Shukri, N. M., 1950. The mineralogy of some Nile sediments. *Quart. J. Geol. Soc. London* **105,** 511–534.

Shukri, N. M., and Azer, N., 1952. The mineralogy

of Pliocene and more recent sediments in the Faiyum. *Inst. Desert Egypte Bull.* 2(1), 10–39.

Sieberg, A., 1932. *Erdbebengeographie, Handbuch der Geophysik.* Berlin, IV, 687–1006.

Simons, E. L., and Wood, Albert E., 1968. Early Cenozoic mammalian faunas, Faiyum Province, Egypt. *Peabody Mus. Nat. Hist., Yale Univ. Bull.* **28**, 1–105.

Smith, P. E. L., 1976. Stone-age man on the Nile. *Scientific Amer.* **235**, 30–38.

Sneh, A., and Weissbrod, T., 1973. Nile Delta: The defunct Pelusiac branch identified. *Science* **1880**, 59–61.

Sneh, A., Weissbrod, T., and Perath, I., 1975. Evidence from an Ancient Egyptian frontier canal. *Amer. Scientist* **63**, 542–548.

Sonnenfeldt, P., 1974. The Upper Miocene evaporite basins in the Mediterranean region—a study in paleo-oceanography. *Geol. Rundschau* **63**, 1133–1172.

Squyres, C. H., and Bradley, W., 1964. Notes on the Western Desert of Egypt, *Conf. Petrol. Expl. Soc., Libya,* 99–105.

Stanley, D. J., 1977. Post-Miocene depositional patterns and structural displacement in the Mediterranean. In: *The Ocean Basins and Margins,* 4A Nair, E. M., Kanes, W. H., and Stehli, G., (eds.). Plenum, New York, 77–150.

Stanley, D. H., and Maldonado, A., 1977. Nile cone: Late Quaternary stratigraphy and sediment dispersal. *Nature* **266**, 129–135.

Stromer von Reichenbach, E., 1902a. Wirbeltierreste aus dem mittleren Pliocaen des Natrontales und rezente Saeugetierreste aus Aegypten. *Zeitschr. D. Geol. Ges. (Berlin)* **3**, 108–116.

Stromer von Reichenbach, E., 1902b. Die altertertiaeren Saeugetiere des Fajum. *Naturwiss. Wochenschr. (Berlin)* **18**, 145–147.

Stromer von Reichenbach, E., 1905. Fossile Wirbeltier-Reste aus dem Uadi Faregh und Uadi Natrun in Aegypten. *Abh. Senckenb. Naturf. Ges. (Frankfurt, a.M.),* **2**, 99–132.

Stromer von Reichenbach, E., 1907. Geologische Beobachtungen im Fajum und am unteren Niltale in Aegypten. *Abh. Senckenb. Naturf. Ges. (Frankfurt, a.M.)* **3**, 133–147.

Summerhayes, C. P., Sestini, G., Misdorp, R., and Marks, N., 1978. Nile Delta: Nature and evolution of continental shelf sediments. *Marine Geol.* **27**, 42–65.

Szabo, B. J., and Butzer, K. W., 1979. Uranium-series dating of lacustrine limestones from pan deposits with Final Acheulian assemblage at Rooi-

dam, Kimberly District, South Africa. *Quaternary Res.* **11**, 257–260.

Thiede, J., 1978. A glacial Mediterranean. *Nature* **276**, 680–683.

Thunnell, R. C., Williams, D. F., and Kennett, J. P., 1977. Late Quaternary Pleoclimatology, stratigraphy and sapropel history in Eastern Mediterranean deep-sea sediments. *Marine Micropaleontology* **2**, 371–388.

Tobia, S. K., and Sayre, E. V., 1974. *An analytical comparison of various Egyptian soils, clays, shales and some ancient pottery by neutron activation.* Report Ain Shams University Chemistry Dept., Cairo, 20 pp.

Tousson, O., 1923. *Anciennes branches du Nil.* Inst. Egypte Mem. **4**, 212 pp.

Venkatarathnam, K., and Ryan, W. B. F., 1971. Dispersal patterns of clay minerals in the sediments of the Eastern Mediterranean Sea. *Marine Geol.* **11**, 261–262.

Verrmeersch, P., 1970. Une Nouvelle Industrie Epipaleolithique a el Kab en Haute Egypte. *Chronique d'Egypte* **45**, 45–67.

Viotti, C., and Mansour, A., 1969. Tertiary planktonic foraminiferal zonation from the Nile Delta, Egypt U.A.R., Part I: Miocene planktonic foraminiferal zonation. *Proc. 3rd African Micropal. Colloq.,* Cairo, 425–460.

Vita-Finzi, C., 1972. Supply of fluvial sediment to the Mediterranean during the last 20,000 years. In: *The Mediterranean Sea: A Natural Sedimentation Laboratory.* Stanley, J. D. (ed.). Dowden, Hutchinson and Ross, Stroudsburg, Pennsylvania, 43–46.

Vondra, C. F., 1974. The Upper Eocene transitional and near-shore Qasr el Sagha Formation, Faiyum Depression, Egypt, U.A.R. *Geol. Surv. Egypt Annals,* **4**, 79–96.

Wendorf, F. (ed.), 1964. *Contributions to the Prehistory of Nubia.* Southern Methodist University Press, Dallas, Texas.

Wendorf, F. (ed.), 1968 . The Prehistory of Nubia, 2 vols. Southern Methodist University Press, Dallas, Texas.

Wendorf, F., and Said, R., 1967. Palaeolithic remains in Upper Egypt. *Nature* **215**, 244–247.

Wendorf, F., Said, R., and Schild, R., 1970a. New concepts in Egyptian Prehistory. *Science* **169**, 1161–1171.

Wendorf, F., Said, R., and Schild, R., 1970b. Problems of dating the Late Paleolithic in Egypt. In: *Radiocarbon Variations and Absolute Chronology Proc. Nobel Symp.,* I. U. Olsson (ed.) **12**, 57–79.

Wendorf, F., Said, R., and Schild, R., 1970c. Late Paleolithic sites in Upper Egypt. *Archaeologia Polona* **12,** 19–42.

Wendorf, F., Schild, R., and Haas, H., 1979. A new Radiocarbon chronology for Prehistoric sites in Nubia. *J. Field Archeol.,* **6,** 219–223.

Wendorf, F., and Schild, R., 1976. *Prehistory of the Nile Valley.* Academic Press, New York and London, 404 pp.

Wendorf, F., and Schild, R., 1980. *Prehistory of the Eastern Sahara.* Academic Press, New York and London, 414 pp.

Wendorf, F., *et al.,* 1977. Late Pleistocene and recent climatic changes in the Egyptian Sahara. *Geogr. J.* **143,** 211–234.

Wendt, E., 1966. Two prehistoric archeological sites in Egyptian Nubia. *Postilla* **102,** 1–46.

Whiteman, A. J., 1971. *The Geology of the Sudan Republic.* Clarendon Press, Oxford, 290 pp.

Willcocks, N., 1889. *Egyptian Irrigation.* E. and F. N. Spon, London, 367 pp.

Willcocks, N., 1904. *The Nile in 1904.* E. and F. N. Spon, London, 225 pp.

Willcocks, N., and Craig, J. I., 1913. Egyptian Irrigation (3rd ed.), E. and F. N. Spon, London, 448 pp.

Williams, M. A. J., 1966. Age of alluvial clays in the Western Gezira, Republic of Sudan. *Nature* **211,** 270–271.

Williams, M. A. J., and Adamson, D. A., 1974. Late Pleistocene desiccation along the White Nile. *Nature* **248,** 584–586.

Wong, H. K., Zarudzki, E. F. K., Phillips, J. D., and Giermann, K. F., 1971. Some geophysical profiles in the Eastern Mediterranean. *Geol. Soc. Amer. Bull.* **82,** 91–100.

Woodside, J., and Bowin, C., 1970. Gravity anomalies and inferred crustal structure in the Eastern Mediterranean Sea. *Geol. Soc. Amer. Bull.* **81,** 1107–1122.

Wright, L. D., and Coleman, J. M., 1973. Variations in morphology of major river deltas as functions of ocean wave and river discharge regimes. *Amer. Assoc. Petrol. Geol. Bull.* **57,** 370–398.

Youssef, M. I., 1968. Structural pattern of Egypt and its interpretation. *Amer. Assoc. Petrol. Geol. Bull.* **52,** 601–614.

Index

Additional information from this book *The Geological Evolution of the River Nile;*
ISBN *978-1-4612-5843-8_OSFO1* is available at http://extras.springer.com

EXTRAS ONLINE

Additional information from this book *The Geological Evolution of the River Nile;*
ISBN *978-1-4612-5843-8_OSFO2* is available at http://extras.springer.com

EXTRAS ONLINE

Additional information from this book *The Geological Evolution of the River Nile;* ISBN *978-1-4612-5843-8_OSFO3* is available at http://extras.springer.com

EXTRAS ONLINE

Additional information from this book *The Geological Evolution of the River Nile;* ISBN *978-1-4612-5843-8_OSFO4* is available at http://extras.springer.com

EXTRAS ONLINE

Printed in the United States
by Baker & Taylor Publisher Services